PROCEEDINGS C

RECHARGEABLE ZINC BATTERIES

COMMEMORATING THE 100TH BIRTHDAY OF A.N. FRUMKIN

Edited by

A.J. Salkind
Rutgers University
Piscataway, New Jersey

F.R. McLarnon
Lawrence Berkeley Laboratory
Berkeley, California

V.S. Bagotzky
A.N. Frumkin Institute of Electrochemistry
Moscow, Russia

BATTERY DIVISION AND THE A.N. FRUMKIN INSTITUTE OF ELECTROCHEMISTRY OF THE RUSSIAN ACADEMY OF SCIENCES

Proceedings Volume 95-14

THE ELECTROCHEMICAL SOCIETY, INC., 10 South Main St., Pennington, NJ 08534-2896

Library of Congress Catalog Number: 95-61587

ISBN 1-56677-109-9

Printed in the United States of America

Professor John O'M Bockris (left) and Aleksandr Naumovich Frumkin in the lab at The University of Pennsylvania in 1960.

TABLE OF CONTENTS

Section A - History and Fundamentals

Introductory Remarks - *A.J. Salkind, and F. R. McLarnon* 1

A.N. Frumkin's Contribution to the Development of Batteries 5
A.M. Skundin

Zinc Electrode in Alkaline Electrolyte, 12
James McBreen

Chemical Effects on Gassing of Partially-Discharged Zinc Power in Alkaline Solutions, *Jean-Yves Huot* 22

Structure of Zincate Complexes in Concentrated Aqueous Hydroxide Solutions, *K.I. Pandya, A.E. Russell, J. McBreen, and W.E. O'Grady* 33

Application of Aqueous Simulation Software to Battery Research 36
A. Anderko, J. Berthold, M. Rafal and S. Sanders

Corrosion Studies of Zinc Electrode Current Collector 48
L. Binder, M. Ghaemi, and K.V. Kordesch

Section B - Silver Zinc Batteries

Problems in the Field of Silver-Zinc Storage Batteries 57
V.S. Bagotzky, I.E. Yablokova, G.Z. Kazakevich, and T.A. Safronova

Past, Present and Future of Cellulosic Separator in Silver-Zinc Secondary Cells 68
Harlan L. Lewis, C. Grun, and A.J. Salkind

Development of Silver-Zinc Cells of Improved Cycle Life and Energy Density 84
R. Serenyi, J. Kuklinski, D.C. Williams and A.F. Thompson

Design Parameters and Advances in Silver Zinc Secondary Batteries 98
Charles Grun, H. Lewis, and A.J. Salkind

Section C - MnO2-Zinc and Zinc-Air Batteries

Mercury and Lead-Free Rechargeable Zn/MnO2 Cells 109
Josef Daniel-Ivad, and Jean-Yves Huot

Bonded Rechargeable Zinc Anode
M. Weissenbacher, J.O. Besenhard, and K. Kordesch 121

New Generation of Rechargeable Zinc-Air Batteries 135
S. Muller, F. Holzer, C. Schlatter, C. Comninellis and Otto Haas

Power and Performance Characteristics of Rechargeable Zinc-Air Batteries for Long Run Time Portable Power Applications, *M.N. Golovin, L.A. Tinker, and E.S. Buzzelli* 147

Section D - Nickel-Zinc Batteries

Diffusion Limitations to Cycle Life of Nickel-Zinc Batteries and Effects of Vibration and Rotation, *A.J. Salkind, E. Kantner, and C. Grun* 160

Alkaline Solid Polymer Electrolyte Ni/Zn Secondary Batteries 182
S. Guinot, N. Bouzir, J.F. Penneau and J.F. Fauvarque

Impedance Studies of Roll-Pressed Porous Zinc Electrode for Ni-Zn Battery 195
S.G. Doo, H.B. Lee and J.Y. Oh

Passivation of Zinc Electrode in the Sealed Nickel-Zinc Cells 210
H. Chang, Y.B. Pyun, and M.E. Alexeeva

Novel Electrolyte Additives to Enhance Zinc Electrode Cycle Life 220
N. Dodapaneni and D. Ingersoll

Oxygen Ionization in the Sealed Nickel-Zinc Cells 229
Y.B. Pyun, H. Chang and M.E. Alexeeva

Factors Influencing the Electrochemical Deposition of Nickel Hydroxide Films 241
C.C. Strienz, S. Motupally and J.W. Weidner

Section E - Commemorative Testimonial to Professor Alexander N. Frumkin

The Palladium Medal Award - A Background Discussion 253
A.J. Salkind

The Great Academician - A Reminiscence 255
J. O'M. Bockris

Subject Index 259

FACTS ABOUT THE ELECTROCHEMICAL SOCIETY, INC.

The Electrochemical Society, Inc., is an international, nonprofit, scientific, educational organization founded for the advancement of the theory and practice of electrochemistry, electrothermics, electronics, and allied subjects. The Society was founded in Philadelphia in 1902 and incorporated in 1930. There are currently over 6000 scientists and engineers from more than 60 countries who hold individual membership; the Society is also supported by more than 100 corporations through Patron and Sustaining Memberships.

The Technical activities of the Society are carried on by Divisions and Groups. Local Sections of the Society have been organized in a number of cities and regions.

Major international meetings of the Society are held in the Spring and Fall of each year. At these meetings, the Divisions and Groups hold general sessions and sponsor symposia on specialized subjects.

The Society has an active publications program which includes the following:

JOURNAL OF THE ELECTROCHEMICAL SOCIETY - The JOURNAL is a monthly publication containing technical papers covering basic research and technology of interest in the areas of concern to the Society. Papers submitted for publication are subjected to careful evaluation and review by authorities in the field before acceptance, and high standards are maintained for the technical content of the JOURNAL.

THE ELECTROCHEMICAL SOCIETY INTERFACE - INTERFACE is a quarterly publication containing news, reviews, advertisements, and articles on technical matters of interest to Society Members in a lively, casual format. Also featured in each issue are special pages dedicated to serving the interests of the Society and allowing better communication between Divisions, Groups, and Local Sections.

MEETING ABSTRACTS *(formerly Extended Abstracts)* - Meeting Abstracts of the technical papers presented at the Spring and Fall Meetings of the Society are published in serialized softbound volumes.

PROCEEDINGS VOLUMES - Papers presented in symposia at Society and Topical Meetings are published from time to time as serialized Proceedings Volumes. These provide up-to-date views of specialized topics and frequently offer comprehensive treatment of rapidly developing areas.

MONOGRAPH VOLUMES - The Society has, for a number of years, sponsored the publication of hardbound Monograph Volumes, which provide authoritative accounts of specific topics in electrochemistry, solid state science, and related disciplines.

For more information on these and other Society activities, visit the ECS Home Page at the following address on the World Wide Web:

http://www.electrochem.org.

RECHARGEABLE ZINC SYMPOSIUM

INTRODUCTORY REMARKS

Alvin J. Salkind (1) and Frank R. McLarnon (2)

(1) Rutgers University - College of Engineering
P.O. Box 909, Piscataway, NJ 08855-0909, USA
and
UMDNJ-Robert Wood Johnson Medical School
675 Hoes Lane, Piscataway, NJ 08854, USA

(2) Lawrence Berkeley Laboratory, 90-2024
Berkeley, Ca 94720-00001

Zinc has many outstanding qualities as a material for use in various battery systems, as shown in Table 1. The systems for which Zinc is in commercial use as a negative electrode are shown in Table 2. The commercial value of annual production of these systems was over $10 billion at manufacturers pricing in 1994. Although perhaps 90% of these cells were primary cells, zinc is increasingly being used in secondary battery designs. Recent developments in zinc as a component in cells is listed in Table 3. These developments are leading to an increasing role of zinc as a battery component.

We are pleased to be co-organizers of this symposium with Prof. Vladimir Bagotzky of the Frumkin Institute of Electrochemistry of the Russian Academy of Sciences. The papers include contributions from 21 research laboratories in 8 countries, a measure of the universality of the use and study of zinc electrodes.

As shown in the accompanying copy of a postcard sent to Professor Alexander N. Frumkin in 1958 by some members of the Electrochemical Society attending the Toronto meeting, Frumkin had a very wide acquaintance with the Society and many prominent members. Details of his life and collaborations are given in the papers by Prof. Skundin and in a separate discussion by Prof. John Bockris. In 1928, he spent about a year at various American universities including the University of Wisconsin at Madison, the University of Chicago and Princeton University. At several of these he was offered teaching positions, but always indicated his preference to return to Russia.

TABLE 1 - WHY USE ZINC IN BATTERIES?

- Lowest equilibrium potential among metals stable in aqueous electrolytes --> high cell voltages
- Reversible electrode kinetics --> high battery rate capability
- Low equivalent weight --> high battery specific energy
- Low Cost
- Ease of fabrication into battery electrodes using various methods
- Non-toxic
- Resistant to corrosion in alkaline, neutral and acidic electrolytes
- Relatively low hydrogen formation rate --> safety
- Abundant world-wide reserves

TABLE 2 - RECHARGEABLE ZINC BATTERIES

- Zn/KOH/MNO_2
- ZN/KOH/NiOOH
- ZN/KOH/AgO
- ZN/KOH/air
  - -- Electrically rechargeable
  - -- Mechanically rechargeable
- Zn/KOH/PbO_2
- Zn/NaOh/$Na_3Fe(CN)_6$
- Zn/ZnCl2/Cl_2
- Zn/$ZnBr_2$/Br_2

TABLE 3 - ZINC CELL COMPONENTS: SOME RECENT DEVELOPMENTS

- Low-zinc solubility electrolytes
  - -- Reduce the rate of Zn electrode shape change
  - -- Extend cell life

- New Zn electrode structures
  - -- Foamed-metal substrates
  - -- Particulate beds for mechanically rechargeable cells

- Improved bifunctional air electrodes
  - -- Corrosion-resistant carbon substrates
  - -- Better electrocatalysts

- Novel NiOOH electrode structures
- Better understanding of Zn complex structure in alkaline electrolytes

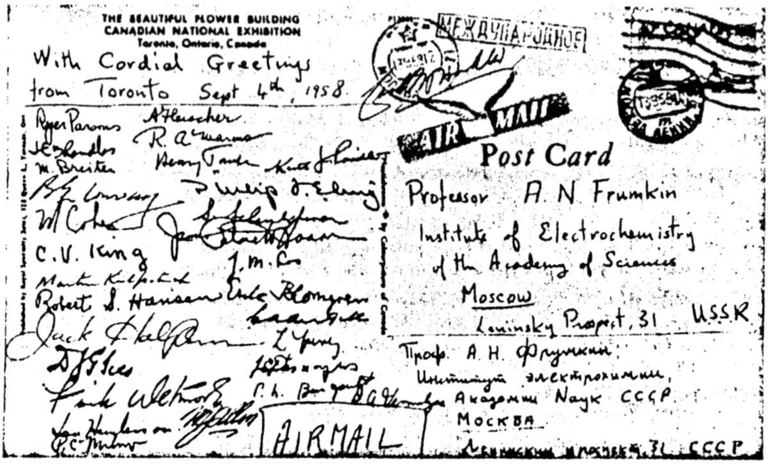

Приветствие А. Н. Фрумкину от иностранных ученых в связи с организацией Института электрохимии АН СССР (1958 г.)

A.N.FRUMKIN'S CONTRIBUTION TO THE DEVELOPMENT OF BATTERIES

A.M.Skundin
A.N.Frumkin Institute of Electrochemistry,
Russian Academy of Sciences
31 Leninsky Prospekt
117071 Moscow, Russia

Alexander N. Frumkin was one of the founders of modern theoretical electrochemistry. At the same time his fundamental investigations, as well as studies of different electrode reactions, had a great influence on the development of various electrochemical devices, in particular, batteries. The most significant contributions of A.N.Frumkin to the development of batteries were: (*i*) solving of the Volta problem, (*ii*) development of a theory of electrochemical kinetics taking into account the influence of electric double layer structure on the rate of electrode processes, (*iii*) an idea that metal corrosion can occur *via* simultaneous anodic and cathodic reactions at the same surface and at the same potential, (*iv*) studies of specific electrode processes, important for batteries, and (*v*) development of the theory of porous electrodes.

When speaking about someone's contribution to the development of batteries, one can consider different levels of such contributions. These can be, *e.g.*, improvements in battery technology or design, or, on the other hand, fundamental discoveries, having a general significance. The contributions of A. N. Frumkin, doubtlessly, belong to the second kind. The list of A. N. Frumkin's publications consists of as many as 775 titles, but among this abundance one can select several fundamental works concerning the philosophy of electrochemistry. These classic investigations had a great influence on the development of electrochemical science as a whole, as well as on the development of practical electrochemical devices, such as batteries.

THE VOLTA PROBLEM

The most important result of Frumkin's early scientific activities was the solution of the Volta problem. This problem arose simultaneously with Volta's discovery and invention of the first galvanic cell. At the initial stage the meaning of the Volta problem was reduced to the question about the origin of the e.m.f. of a galvanic cell. Volta himself considered the e.m.f. as being the result of a contact of two different metals. It must be remembered that all galvanic cells in principle consist of at least two metals (conductors of I kind) and an electrolyte (conductor of II kind). The simplest properly open galvanic circuit can be represented as:

$$M_1 / E / M_2 / M_1 ,$$

where M_1 and M_2 are two different metals, and E is an electrolyte. According to Volta's view, the contact of two metals (*i.e.* M_2/M_1) generates the electric energy of the pile. The insertion of an electrolyte disconnects the other junction of these metals. Volta knew that any circuit, consisting only of metals, has a zero e.m.f. (Volta's law). Thus, Volta was the creator of the "contact" theory of e.m.f. The main arguments in support of this theory were the results of the so-called "Volta basic experiment" - two polished sheets of different metals (*e.g.* copper and zinc) were brought into contact and then disconnected; after disconnection an electroscope showed that zinc is charged positively, and copper is charged negatively.

Of course, this theory does not correspond to the energy conservation law, *i.e.* the theory admitted the existence of a *perpetuum mobile* of the first kind, but this fact didn't trouble Volta. Indeed, the energy conservation law was discovered only 14 years after Volta's death.

At the begining of the XIX century the contact theory of e.m.f. was supported by many scientists, mainly by physicists - Chr. Pfaff, S. Marianini, G. Fechner, G. Zamboni, C.Matteucci, *etc.* In contrast to Volta, some of these scientists admitted that the e.m.f. can arise not only at metal-metal contacts, but also at metal-electrolyte contacts.

Almost simultaneously with the appearance of the contact theory, an alternative "chemical theory" of e.m.f. arose. The pioneer was the Italian scientist G. Fabbroni, who pointed out in 1800 that definite chemical processes (oxidation of metals, formation of salts *etc.*) always occur in experiments with galvanic cells. In 1801 the chemical theory was formulated by N. Gautherot and W. Wollaston. This theory was supported by J. Ritter in Germany, G. Parrot in Russia and by other researchers. Important results confirming the chemical theory were obtained by A. de la Rive. He showed that two metals in different electrolytes can have different polarity. For instance, copper is more negative than tin in salt solutions, but it is more positive in ammonium hydroxide.

Other investigators carried out numerous experiments and confirmed the chemical theory of e.m.f., but nobody could overturn the results of Volta basic experiment. The situation seemed to be hopeless when M.Faraday began his famous experimental researches in electricity.

Faraday performed many witty experiments with cells, made from different metals, different electrolytes *etc.* and he came to the definite conclusion: in all cases when there is no chemical action, the current does not flow, independently of the existence of a contact of two metals, and *vice versa* the current flows always, when there is a chemical action. Faraday studied the influence of the electrolyte nature on the electromotive series of metals (Volta series). He investigated seven different electrolytes and obtained seven different series!

On the whole, Faraday believed the chemical theory to be proved. It is interesting to note that from Faraday's point of view the main argument in favour of this theory was the impossibility to produce electricity from nothing. Faraday didn't know the energy conservation law, but he feeled it. Faraday had an excellent intuition!

The discovery of the energy conservation law struck a blow to the contact theory. Of course, electric energy can not be obtained without the spending of another kind of energy. But the contact theory did not die, because Volta basic experiment could not be refuted.

In the last quarter of the XIX century a new branch of science arose - physical chemistry. Due to works of R.Fauvre, F.-M. Raoult and especially, J. Van't Hoff chemical thermodynamics and the theory of chemical equilibriums were developed. A galvanic cell turned out to be an excellent tool for verifying the Gibbs-Helmholtz equation and the Van't Hoff isotherm. Free energy values of various redox-reactions were obtained from chemical experiments. The e.m.f of galvanic cells based on these reactions were measured. In all cases the measured values were equal to those calculated from free energy. This was the solution of the first part of the Volta problem. The question "What is origin of e.m.f. of a galvanic cell?" was answered.

However, the second part of the Volta problem, *i.e.* the question about localization of the e.m.f. in a cell, was still open. It was clear that the e.m.f. of a cell can be considered as the algebraic sum of potential drops, but the question about the localization of these drops remained. In an attempt to answer this question, W. Nernst in 1887-1889 developed the theory of electrode potentials. According to Nernst, the potential drops reside exclusively at the metal-electrolyte interfaces. In the simplest case, Nernst treated the electrode as a metal in a solution of its salt, and considered the equilibrium between cations in metal and in solution. Using thermodynamic methods, he obtained his famous equation for the concentration dependence of electrode potential (Nernst equation). The constant in this equation is correlated with the position of the metal in the Volta series.

Thus, Nernst was convinced that potential drops at metal-metal junctions are absent. As for the classical Volta's basic experiment, Nernst and other representatives of the new electrochemical school believed that Volta had dealt not with pure metals, but with metal samples, having surface moisture films (even in trace amounts). These films, according to Nernst, played the role of an electrolyte. It seemed that now the Volta problem was solved.

Two decades later, I. Langmuir repeated Volta's basic experiment, but using a high temperature vacuum treatment of metals. At these conditions the metal surfaces were free from any moisture and other contaminations. And inspite of this, Langmuir in 1916 obtained results, identical to Volta's ones!

Taking into account these results, Langmuir renewed the contact theory. According to Langmuir, the e.m.f. is localized at the metal-metal contacts. If the electrolyte in the cell contains appropriate ions and there is the possibility of electrode reactions, the cell can produce electricity. If such ions are absent, the electrode are polarized and current does not flow through the system.

After Langmuir's work a "strange situation" arose in battery science. Chemists were concerned with galvanic cells and usually ignored contact potentials; physicists, on the contrary, dealt with contact potentials, work functions *etc.*, and considered galvanic cells as very complex, complicated, and, therefore, uninteresting systems.

Just when Langmuir finished his work on contact potentials in electrochemical circuits, young Frumkin began his studies on electrocapillarity. In 1919 he presented a thesis "Electrocapillary phenomena and electrode potentials", in which a new idea was formulated. In contrary to Nernst, Frumkin concentrated his attention on ideally polarizable electrodes. Electrocapillary curves give a possibility to measure zero charge potentials of such electrodes. According to Nernst, the absolute zero charge potentials of any metal is equal to zero, and therefore the zero charge potentials of all metals, measured *vs.* the same reference electrode, must be the same. Frumkin pointed out that zero charge potentials of different metals are different. Later, in 1928 he drew a conclusion that the difference between the zero charge potentials of two metals is approximately equal to the difference of their work functions, *i.e.* is approximately equal to the contact potential. This idea was the key for solving the Volta problem.

Frumkin postulated that all interfaces in a galvanic circuit contribute to the e.m.f. In the special case, the circuit consists of two electrodes having their zero charge potentials. The corresponding concentrations of the potential-determining ions can be calculated from the Nernst equation. In this case the e.m.f is equal to the contact potential. If the concentrations of the above mentioned ions are different, the potential drops in the ionic double layers at the metal-solution interfaces change according to the Nernst equation, and the e.m.f. value changes also.

Thus, Frumkin combined two views, which seemed to be "incompatible".

KINETICS OF ELECTRODE PROCESSES

Performances of batteries to a great extent depend on the kinetics of electrode processes. It is clear that useful (current-producing) processes must be as fast as possible, whereas various side reactions, resulting in self-discharge and efficiency diminishing, must be inhibited.

A. Frumkin began his researches in electrochemical kinetics in the early thirties. Actually, at that time electrochemical kinetics was not yet developed, though numerous papers had been published, in which the dependence of electrochemical reaction rates on various conditions (including the potential) was studied. In 1905 the famous Tafel equation - the main empirical equation of electrochemical kinetics - was formulated.

The Tafel equation establishes a simple correlation between current density i and overpotential η for the reaction of cathodic hydrogen evolution:

$$\eta = a + b\log i. \qquad [1]$$

J. Tafel himself tried to understand the meaning of this equation and he suggested the first theory of a hydrogen overvoltage. The whole process of hydrogen evolution consists of several consecutive steps: (i) approach of hydrogen ions to the electrode surface, (ii) ion discharge with formation of adsorbed hydrogen atoms, (iii) recombination of two hydrogen atoms to a molecule, (iv) formation of bubble nuclei, (v) bubbles growth and their withdrawal. According to Tafel, step (iii) is the slowest and, therefore, the rate-determining step. On the whole, it was believed that any step can be in principle rate-determining, except step (ii) - the actual electrochemical step of charge transfer through the interphase. The charge transfer was considered as an infinitely fast process.

The hydrogen evolution reaction became the most popular and the most important one in subsequent kinetic investigations due to its apparent simplicity and its significance in practice (in the electrochemical industry, in batteries, in corrosion processes *etc.*).

Later it was found that the Tafel equation has a very general importance: not only the kinetics of hydrogen evolution, but also the kinetics of most other electrochemical reactions obey this equation. However, the Tafel theory contradicted to experimental results. Firstly, the recombination theory predicts the value of the constant "b" to be equal to 2.3RT/2F, *i.e.* about 30 mV per decade; in practice b for many cases is as much as 120 mV. Secondly, the theory couldn't explain the substantial effect of the pH value (*i.e.* hydrogen ion concentration) on the rate of hydrogen evolution. Since hydrogen ions are not involved in the recombination step, the rate of this step should not depend on the solution pH, whereas in many cases the rate of hydrogen evolution is proportional to the hydrogen ions concentration. And finally, the Tafel equation is valid for processes without recombination steps, *e.g.* for oxidation of ions Mn^{2+} and for reduction of ions Mn^{3+}.

In 1930 M. Volmer and T. Erdey-Gruz suggested a new theory of hydrogen evolution, in which step (ii) was considered to be the slowest one. They obtained the following dependence of i on potential E:

$$i = k[H^+]\exp(-\alpha FE/RT) \qquad [2]$$

The constants in this equation have a clear meaning: k is the rate constant, α is the so-called transfer coefficient, which reflects the shares of the overall potential drop influencing the rate of the direct and the reverse reactions. The value of α must be between 0 and 1; so in many cases α is approximately 0.5. One can see that equation [2] is easily converted into the Tafel equation, taking into account that $\eta = E - E_{reversible}$, and $b = 2.3RT/\alpha F$. If $\alpha = 0.5$, then b = 120 mV in accordance with experimental values.

The theory of Volmer and Erdey-Gruz was able to overcome some difficulties of Tafel's recombination theory, but nevertheless there were some discrepancies with experiments. For instance, it was known that salt additives to acid solutions resulted in current density changes even at constant values of hydrogen ion concentration. This fact could not be explained by equation [2].

A. Frumkin pointed out that Volmer and Erdey-Gruz did not take into account the obvious fact that the discharge of hydrogen ions, *i.e.* the elementary act of the electrochemical process, always occurs within the electric double layer. In his classical work in 1933 Frumkin treated the influence of the double layer on the kinetics of hydrogen ion discharge. Volmer and Erdey-Gruz believed the surface concentration of hydrogen ions $[H^+]_s$ being equal to their bulk concentration $[H^+]$. Meanwhile Frumkin used for the value of $[H^+]_s$ an expression from the double layer theory of Gouy-Chapman-Stern:

$$[H^+]_s = [H^+]\exp(-\psi_1 E/RT), \qquad [3]$$

where ψ_1 is the potential drop in the double layer, corresponding to its diffuse part.

At the same time, during discharge an electron overcomes not the whole potential drop E, but only $E - \psi_1$, *i.e.* a fraction corresponding to the dense part of the double layer. Thus, equation [2] converts into

$$i = k[H^+]\exp(-\psi_1 F/RT)\exp[-\alpha F(E-\psi_1)/RT] \qquad [4]$$

Such a conversion of equation [2] into [4] was called the "Frumkin correction". This name must not diminish the significance of the Frumkin's work of 1933. Actually it was not a small improvement of the equation, but it was the ideology of the modern electrochemical kinetics. A lot of papers after the pioneer Frumkin's work were devoted to the investigation of the influence of the double layer and of the interface structure on the kinetics of electrode processes.

It must be mentioned that in 1952 a monograph "Kinetics of electrode processes", written by Frumkin and his pupils, was published in the USSR. It was the first book of this topic in the world literature.

METAL CORROSION

Corrosion of metals is a very important process for batteries because it results in self-discharge. The electrochemical nature of corrosion was discovered as early as in 1813 by H.Davy. In 1830 A. de la Rive suggested his theory of local corrosion microcells. According to this theory a metal surface in the most cases is contaminated by microinclusions of more noble materials. For instance, the surface of steel can be contaminated by the inclusions of carbides *etc.* Therefore, the contaminated metal acts as a number of short-circuited cells, in which the metal itself plays the role of an anode, and the inclusions play the role of cathodes. Anodic processes (metal dissolution) occur at the metal surface, and cathodic ones (*e.g.* hydrogen evolution or dissolved oxygen reduction) occur at the surface of the inclusions. This approach was commonly accepted up to thirties of the XX century.

In 1932 A.Frumkin came to the conclusion that the existence of different anodic and cathodic sites at the metal surface is not a necessary condition for electrochemical corrosion. It is interesting to note, that this conclusion was not an object of any special investigation, but was only a footnote in a paper, devoted to an analysis of the meaning of the transfer coefficient α in equation [2]. Frumkin pointed out that the anodic process (metal dissolution) and the cathodic one (hydrogen evolution) can occur at the same surface and at the same potential even at metals with an absolutely homogeneous surface, *e.g.* at liquid amalgams. This idea was suggested also by L. Hammett and A. Lorch independently of Frumkin. Six years later C.Wagner and W.Traud expanded this idea for all cases of electrochemical corrosion and formulated the principle of mixed (compromise) potentials - the main principle of the modern corrosion theory.

Thus, in the thirties a quantitative theory of self-dissolution of metals, both with and without surface contaminations, was developed. Corrosion phenomena were treated on the base of electrochemical kinetics. For the case of solid metal corrosion (nickel and lead) this theory was experimentally proved by A.Frumkin *et al.* in 1941-1951.

ANODIC OXIDATION OF ORGANIC SUBSTANCES

At the very beginning of the sixties great efforts were made in attempts to develop fuel cells. The idea of fuel cell was formulated by W. Ostwald yet in 1894, but large-scale investigations were initiated only almost sixty years later. In particular, many investigations were aimed to the development of fuel cells with liquid organic fuels, such as alcohols, hydrocarbons *etc.* A.N.Frumkin was a great enthusiast of the fuel cell concept, and he carried out an extensive work to organize such investigations in the former USSR.

Frumkin understood that successes in the development of fuel cells to a great extent depend on researches in the field of kinetics and mechanism of electrode processes.

Frumkin and his pupils studied the mechanism of anodic oxidation of simple organic substances - methanol, other low molecular alcohols, aldehydes, formic acid *etc.* They showed that adsorption of these compounds at platinum group metals occurs *via* deep destruction of the adsorbed molecules. In particular, this destruction is due to dehydrogenation, therefore the electrode surface in a solution of organic compounds is covered by both adsorbed hydrogen atoms and carbon-containing fragments. So, the oxidation of low molecular organic compounds is actually an oxidation of these adsorbed species. This oxidative step can proceed either by an electrochemical, or by a chemical mechanism.

Due to various reasons, the success in the development of practical fuel cells with liquid organic fuels was rather modest, in particular in comparison with the development of hydrogen-oxygen fuel cells. However, the investigations of adsorption and oxidation of organic substances at electrodes resulted in the creation of a new field of electrochemistry - electrocatalysis. (It must be stressed that interest to fuel cell with liquid organic fuels is renewed in our days).

THEORY OF POROUS ELECTRODES

Electrodes in most kinds of batteries have a porous structure. As a rule, they represent two-phase systems - porous bodies with electronic conductance filled with a liquid electrolyte. Electrode processes in such systems (systems with distributed parameters) are complicated by transport difficulties - slow diffusion rate, ohmic losses, *etc.* .

The true internal surface area of porous electrodes is much greater than their apparent geometric surface area. This enables one to obtain high overall currents with comparatively low current densities and, therefore, low polarization. Meanwhile, in practice the overall current at porous electrodes is less than that, calculated from the true internal surface area and from the current density at a smooth electrode under the same conditions. In other words porous electrodes have a relatively low efficiency. The loss of efficiency is due to the fact that different sites within the electrode are not equally accessible to the electrode reaction. Concentration gradients and ohmic drops are possible in the electrolyte in the pores. Hence, the local current densities and, therefore, local potential values are different at different depths of the porous electrode.

A.N.Frumkin was one of the first electrochemists, who tried to carry out the difficult task of a mathematical description of these systems. As early as in 1949 he published a paper "On the corrosion process distribution upon tube length". It was a very fruitful approach to consider a single cylindrical pore as a simple model of very complex real porous electrodes. Frumkin obtained strict equations for the potential distribution for two cases of polarization - linear and exponential. It was important that Frumkin defined the limits of the one-dimensional approach.

Later the theory of porous electrodes was intensively developed, in particular, by representatives of Frumkin scientific school.

ZINC ELECTRODE IN ALKALINE ELECTROLYTE

James McBreen
Department of Applied Science
Brookhaven National Laboratory
Upton, NY 11973

The zinc electrode in alkaline electrolyte is unusual in that supersaturated zincate solutions can form during discharge and spongy or mossy zinc deposits can form on charge at low overvoltages. The effect of additives on regular pasted ZnO electrodes and calcium zincate electrodes is discussed. The paper also reports on *in situ* x-ray absorption (XAS) results on mossy zinc deposits.

INTRODUCTION

There are several reviews of the zinc electrode in alkaline electrolyte (1, 2). Recent fundamental work on the zinc electrode in alkaline electrolyte has focused on methods for reducing the solubility of ZnO, the mechanisms of formation and decomposition of supersaturated zincate electrolytes, the growth of mossy deposits on charge at low current densities, and the effect of additives and pulsed charging on the morphology of the zinc electrode.

Many of the problems of rechargeable zinc batteries are due to zinc electrode shape change and zinc dendritic growth. These problems can be traced to the high solubility of the discharge products of the zinc electrode. In the case of the silver-zinc battery there are very limited options for dealing with these problems because of the necessity of using ≈45% KOH to suppress the solubility of Ag_2O. Sacrificial cellulosic separators are also necessary to inhibit migration of silver species to the zinc electrode. With the use of nickel, manganese dioxide or air positive electrodes there are many more options for the choice of separator and the electrolyte composition. This has led to many design changes and electrolyte formulations that improve the cycle life of the zinc electrode. The design changes can be classified as soluble zinc electrodes or zinc electrodes of limited solubility

ZINC ELECTRODES

Soluble zinc electrodes: The equilibrium solubility of ZnO in KOH depends on the KOH concentration and increases from 6 g/l in 10% KOH to 53 g/l in 30% KOH. However, when a zinc electrode is discharged in an alkaline electrolyte supersaturated zincate solutions can be formed. The zincate solubility can be as much as three times the equilibrium solubility. These solutions, however, are not supersaturated in the ordinary sense of the term. Neither seeding or shock causes precipitation of ZnO. Instead, the ZnO is precipitated slowly in a manner resembling a decomposition reaction. The presence of Li^+, $SiO_2^=$ or sorbitol retards the process. Depending on conditions, it can take anything from hundreds of hours to a year to reach the equilibrium solubility (3). Developers, such as Sorapec, have taken advantage of this and have developed nickel zinc batteries with the discharged product of the zinc electrode completely dissolved in the electrolyte (4).

Electrodes of limited solubility: The other approach to minimizing problems at the zinc electrode is to reformulate the electrode or electrolyte composition so as to reduce the ZnO solubility. This can be done either by incorporating $Ca(OH)_2$ in the zinc electrodes or by use of electrolytes with minimal hydroxyl content containing highly soluble salts such as phosphate, borate carbonate or fluoride. The benefits of these approaches have been reviewed by Bass *et al.*(5).

Effect of additives: The two approaches for the zinc electrode, outlined above, increase the cycle life considerably if they are combined with the use of additives and/or special charging techniques. In conventional pasted zinc electrodes small amounts of additives such as HgO or PbO can have enormous effects on the zinc electrode. Additives such as HgO or Ga_2O_3 accelerate shape change and decrease cycle life, whereas additives such as Pb or mixtures of PbO and CdO are beneficial. Observations at Brookhaven National Laboratory (BNL) during the charging of pasted ZnO electrodes showed that beneficial additives promoted zinc deposition in the interior of the electrode in the vicinity of the grid current collector, whereas additives such as HgO or Ga_2O_3 promoted zinc deposition on the front of the electrode that abuts the separator. Additives such as PbO and $In(OH)_3$ promote the formation of dense zinc deposits that increase the polarizability of the electrode. HgO and Ga_2O_3 on the other hand promote finely divide zinc deposits and electrodes with low polarizability and high rate capability.

Additives in calcium zincate electrodes: Additives such as PbO are particularly helpful in increasing the discharge capacity of electrodes of low solubility containing $Ca(OH)_2$. Electrolytes of low solubility can lead to passivation of the zinc electrode on discharge. The effect of additives on the performance of calcium zincate electrodes has been examined at BNL.

The effect of additives on zinc utilization was determined by tests in 1 Ah nickel-zinc cells with two electrodes. The zinc electrodes (7.62 x 5.08 cm) were prepared by mixing 3 g of PTFE bonded ZnO with 3 g of $Ca(OH)_2$ and an additive and pressing the mix on to a silver grid. The cells were wet down with 20% KOH containing 11 g LiOH + 30 g ZnO/l. After soaking overnight the cells were charged at 75 mA for 17 h. The zinc utilization was determined by discharging the cells at 500 mA (C/2 rate) till the zinc electrode reached -1.0 V vs. a Hg/HgO reference. The results are shown in Fig. 1. It is clear that additives can have a beneficial or detrimental effect on the zinc utilization in zinc electrodes with limited solubility. Additions of PbO and Cu powder are especially helpful.

Additives in pasted ZnO electrodes: Work at BNL (6-8) and the work of Himy, Wagner (9) and others indicate that additives, either added singly or in combination, can have an enormous effects on the rate of electrode shape change and the polarizability of the electrode. Additives such as HgO and Ga_2O_3 decrease the electrode polarizability and accelerate shape change. Additives like PbO and $In(OH)_3$ increase the electrode polarizability and inhibit shape change. The addition of CdO has a minimal effect on electrode polarizability. The best compromise between cycle life and high rate performance is to add combinations of additives such as PbO + CdO.

Mechanism of the effect of additives: The mechanism by which additives work in pasted zinc electrodes is not fully understood. The work of McBreen and Gannon indicates that additives do affect the zinc morphology and the distribution of zinc within the pores of the electrode and across the surface of the electrode (6-8). One unexpected finding on additives in pasted ZnO electrodes was that on the first charge the oxide additives were often quantitatively deposited as the metal prior to the onset of zinc deposition (6). This was true for both HgO and Tl_2O_3. It is difficult to envision how an additive of limited solubility can be quantitatively reduced in a non-conducting matrix such as ZnO. The fact that this occurs and that the effect of the additive persists throughout cycling indicates that the additive effect may be a substrate effect that controls zinc nucleation and morphology. The effect of additives on zinc utilization in electrodes of limited solubility is even more mysterious. The effect of the Cu powder can be rationalized by the fact that Cu substrates promote the growth of finely divided deposits that inhibit passivation on discharge. However, PbO is known to promote dense deposits so a decrease in zinc utilization would be expected.

MOSSY ZINC DEPOSITS

Mechanisms for formation of mossy zinc deposits: Zinc deposition from alkaline zincate electrolytes is unusual in that finely divided spongy or mossy deposits are formed at low current densities (10). In the case of other metals finely divided deposits are only formed at very high current densities (11). Three types of mechanisms have been proposed to explain the anomalies in zinc deposition. These are:

1. Zinc deposition through surface oxide films (12-14).

2. Instabilities due to multiple steady states caused by autocatalytic processes (15, 16).

3. Stabilization of dendrites and superlattices by codeposition of adsorbates such as hydrogen (17).

Wiart and co-workers have proposed that at low overvoltages zinc deposition occurs through a non stoichiometric ZnO film on the zinc surface (12-14). This is illustrated in Fig. 2. This model was developed to account for the polarization curves, ac impedance measurements, and morphological features in scanning electron micrographs. To explain all the observed phenomena they also postulated the following. Parts of the film are insulating and block the surface; the remainder has both ionic and electronic conductivity which permits deposition. The thickness of the film can be affected by potential, Pb additives and the presence of anodically generated zincate species. The latter promote the growth of spongy zinc by decreasing the film thickness.

Epelboin *et al.* proposed the following reaction mechanism to explain S shaped polarization curves and inductive loops in ac impedance curves, during zinc deposition (15). The reaction scheme involves an autocatalytic step. This reaction scheme has been slightly modified by Lee and Jorné (16) and is as follows:

$$H^+ + e^- \rightarrow H_{ad}$$

$$H^+ + H_{ad} + e^- \rightarrow H_2$$

$$Zn^{2+} + Zn_{ad}^+ + e^- \rightarrow 2Zn_{ad}^+$$

$$Zn_{ad}^+ + H_{ad} \rightarrow H^+ + Zn$$

$$Zn_{ad}^+ + e^- \rightarrow Zn$$

$$Zn^{2+} + H_{ad} \rightarrow Zn_{ad}^+ + H^+$$

The autocatalytic step can yield multiple steady states and instabilities. It can also induce an inverse relationship between the rate of zinc deposition and zinc ion concentration. Epelboin postulated that these instabilities were the cause of mossy deposits.

Clarke and co-workers have reported on a very interesting *in situ* high resolution x-ray diffraction study of dendritic zinc deposition from a $ZnSO_4$ electrolyte (17). The x-ray diffraction data displayed extra peaks in addition to the strongest expected zinc reflections. The spacing and symmetry of the additional peaks indicated a $\sqrt{3}\times\sqrt{3}$ superlattice ordering in the basal plane of zinc. They proposed a "keystone" model to explain the formation of superlattices. The model is illustrated in Fig. 3. The model depends on the codeposition of an adsorbate which interferes with normal epitaxial deposition. They postulate that the most likely adsorbate is hydrogen. This is supported by pyrolytic analysis of the deposit which indicates a hydrogen content as large as 1.7%

by weight. This exceeds the hydrogen content of $FeTiH_2$. A similar mechanism may result in moss formation.

XAS studies of deposition of mossy zinc: The zinc deposition studies were carried out in the cell shown in Fig. 4. XAS spectra were recorded for the zincate electrolyte, prior to zinc deposition. Zinc was deposited under potentiostatic conditions (-45 mV or -65 mV vs. a Zn wire reference) in a thin cavity (~0.5 mm). After sufficient zinc was deposited to fill the cavity above the level of the x-ray window, transmission XAS measurements were made, with the electrode still under potential control.

Figure 5 shows a comparison of the Fourier transform of the EXAFS for a zinc foil and a mossy zinc deposit at -65 mV. There is no evidence for a Zn-O contribution at 1.5 Å. This indicates that the electrolyte within the electrode pores is completely exhausted of zincate and is essentially a pure KOH solution. Also there is no evidence for significant adsorbed oxide on the zinc deposit. The first Zn-Zn peak for the mossy deposit is much lower than that found for the metal foil. The data for this peak could be fitted to a single Zn-Zn coordination shell with a coordination number of 4.09 and a bond distance of 2,63 Å. The nearest coordination shell for bulk zinc has 6 atoms at 2.67 Å and 6 others at 2.89 Å. If the zinc deposit was oriented with the c-axis parallel to the lines of current, then only the nearest neighbors in the basal plane would be observed; that is 6 atoms at 2.67 Å. This is due to the polarization of the x-ray beam in the plane of the synchrotron ring. Such effects have actually been reported for single crystal zinc (18). This could partially explain the result. However, the calculated coordination number is considerably lower than six. Clarke, on the basis of x-ray diffraction results has reported that mossy zinc is a nanophase material with a crystallite size of ≈20 Å (25). The very low coordination number would be consistent with this. All of this indicates that mossy zinc is a very unusual material.

Figure 6 shows Fourier transforms of the EXAFS for zinc moss at -45 mV, zinc foil, and 8.4 M KOH + 0.74 M KOH. At -45 mV there is a large Zn-O contribution. This is either due to the presence of zincate in the electrode pores or the presence of the Zn_xO_y films, postulated by Wiart and co-workers (12-14). Because of the presence of two types of zinc species qualitative analysis of the spectra was impossible.

CONCLUSIONS

The work discussed above indicates that the electrochemistry of zinc deposition in alkaline electrolyte is exceedingly complex. The same is true for the discharge process. The formation and stability of the supersaturated zincate solutions remains an enigma. The work of Clarke, a physicist, has revealed complexities in zinc moss deposition that were hitherto unknown. All of this indicates the need for an interdisciplinary approach to the study of the zinc electrode. There are apparently autocatalytic processes involved in both the deposition process on charge and the precipitation of ZnO on discharge. This means that on charge zinc prefers to deposit at sites where there is zinc and on discharge ZnO precipitates preferentially at sites where there is ZnO. This could account for some of the instability of the electrode on cycling.

ACKNOWLEDGMENTS

The author gratefully acknowledges support of the U.S. Department of Energy, Division of Materials Sciences, under Contract Number DE-FG05-89ER45384 for its role in development and operation of Beam Line X-11 at the National Synchrotron Light Source (NSLS). The NSLS is supported by the Department of Energy, Division of Materials Sciences under Contract Number DE-AC02-76CH00016. The experimental work was supported by the Assistant Secretary for Energy Efficiency and Renewable Energy, Office of Transportation Technologies, Electric and Hybrid Propulsion Division, USDOE under Contract Number DE-AC02-76CH00016.

REFERENCES

1. H. André, U. S. Pat. 2,317,711 (1943).

2. J. McBreen, in Power Sources for Electric Vehicles, B. D. McNicol and D. A. J. Rand, Editors, Elsevier, Amsterdam (1984), pp. 541-572.

3. C. Debiemme-Chouvy, J. Vedel, M.-C. Bellisent-Funel and R. Cortes, J. Electrochem. Soc., 142, 1359 (1995).

4. G. Bronoel, A. Millot and N. Tassin, J. Power Sources, 34, 243 (1991).

5. K. Bass, P. J. Mitchell, G. D. Wilcox and J. Smith, J. Power Sources, 35, 333 (1991).

6. J. McBreen and E. Gannon, Electrochim. Acta, 26, 1439 (1981).

7. J. McBreen and E. Gannon, J. Electrochem. Soc., 130, 1980 (1983).

8. J. McBreen and E. Gannon, J. Power Sources, 15, 169 (1985).

9. A. Himy and O. C. Wagner, U. S. Pat. 4,084,047.

10. S. Arouete, K. F. Blurton and H. G. Oswin, J. Electrochem. Soc., 116, 166 (1969).

11. N. Ibl, Advances in Electrochemistry & Electrochemical Engineering, Vol. 2, P. Delahay and C. W. Tobias, eds., Wiley, New York (1962) p. 49.

12. C. Cachet, S. Saïdani and R. Wiart, Electrochim. Acta, 33, 405 (1988).

13. C. Cachet, S. Saïdani and R. Wiart, Electrochim. Acta, 34, 1249 (1989).

14. C. Cachet, S. Saïdani and R. Wiart, J. Electrochem. Soc., 138, 678 (1991).

15. I. Epelboin, M. Ksouri, E. Lejay and R. Wiart, J. Electrochem. Soc., 122, 1206 (1975).

16. M. G. Lee and J. Jorné, J. Electrochem. Soc., 139, 2841 (1992).

17. D. G. Grier, K. Allen, R. S. Goldman, L. M. Sander and R. Clarke, Phys. Rev. Lett., 65, 2152 (1990).

18. G. S. Brown, P. Eisenberger and P. Schmidt, Solid State Comm., 24, 201(1977).

19. R. Clarke, private communication.

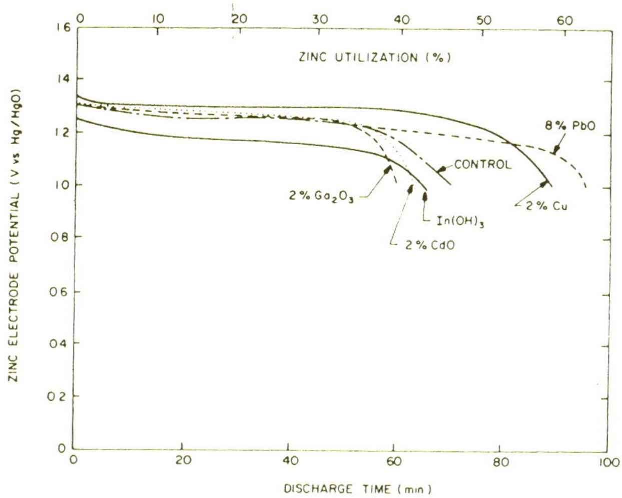

Fig. 1. Zinc utilization for calcium zincate electrodes with various additives. The control had no additive. The additions were 2% for $Ga(OH)_2$ CdO, $In(OH)_3$, and Cu. The PbO addition was 8%.

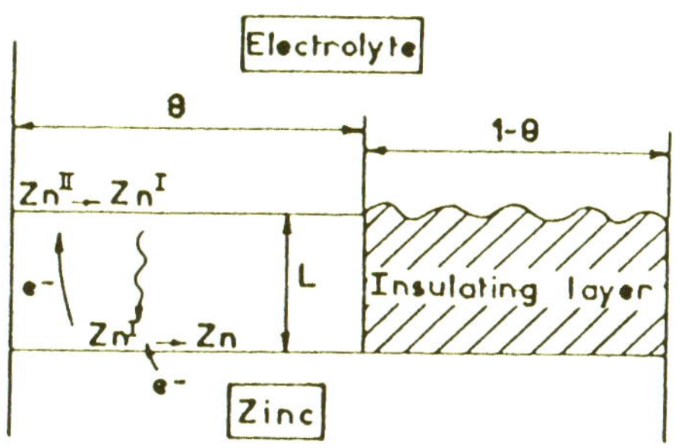

Fig. 2. Model for zinc deposition through an oxide film at low overvoltages (14).

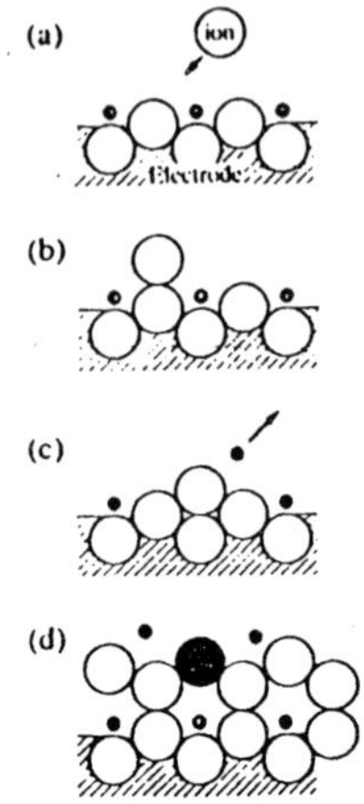

Fig. 3. The "keystone" model to explain formation of superlattice structures (17). The large circles indicate zinc, both before and after incorporation into the lattice. The small circles indicate an adsorbate such as hydrogen. Without interference from the adsorbate the result is compact epitaxial deposition, as shown in (c). With adsorption of hydrogen at the energetically favorable sites, deposition can occur at other sites such as in (b). This type of deposition and the deposition of "keystone" atoms (the dark circle) (see (d)) produces a $\sqrt{3}\times\sqrt{3}$ superlattice when the growth is in the [110] direction.

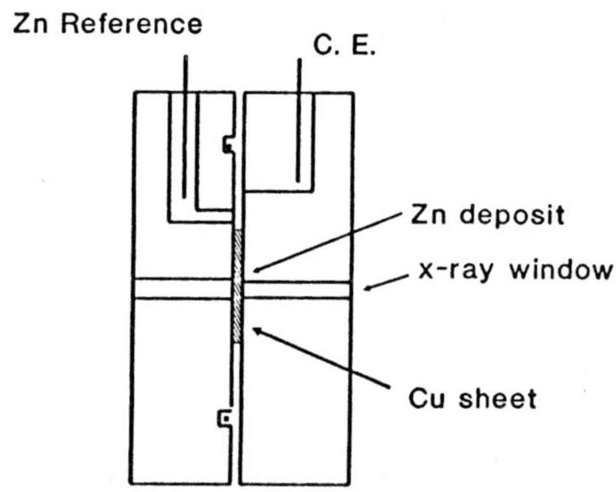

Fig. 4. Cell for *in situ* XAS studies of deposition of zinc moss.

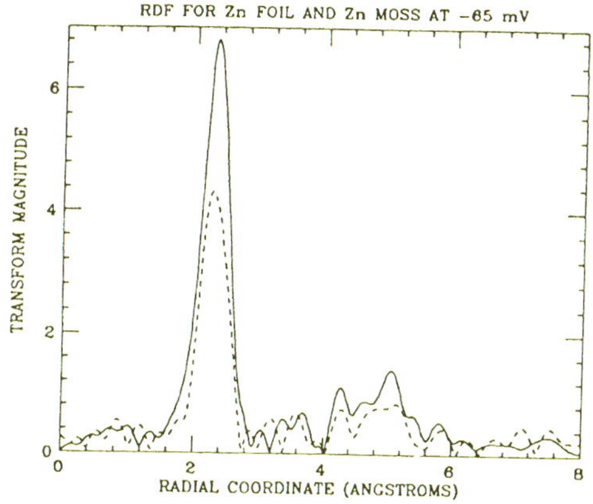

Fig. 5. Fourier transforms of the EXAFS for zinc foil (———) and zinc moss at -65 mV (- - -).

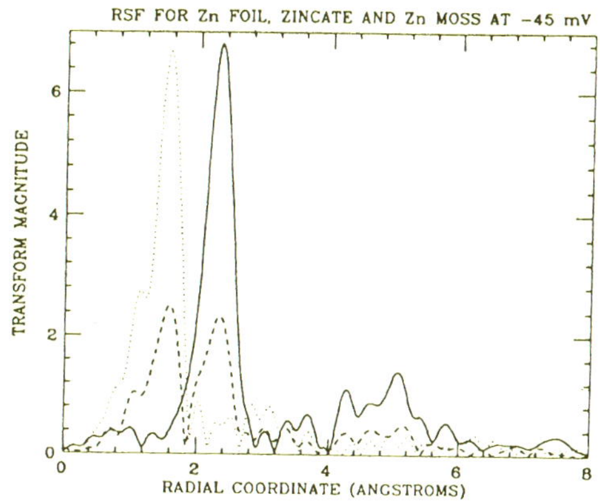

Fig. 6. Fourier transforms of the EXAFS for zinc foil (———), zinc moss at -45 mV (- - -) and an 8.4 M KOH + 0.74 M ZnO electrolyte (···).

CHEMICAL EFFECTS ON GASSING OF PARTIALLY-DISCHARGED ZINC POWDER IN ALKALINE SOLUTIONS

Jean-Yves Huot
Noranda Advanced Materials, c/o Centre de technologie Noranda,
240 Hymus Blvd., Pointe-Claire, Québec, Canada H9R 1G5

Gassing of partially-discharged zinc powder in alkaline solutions was evaluated. It was shown that after-discharge gassing depends on anodic discharge conditions, chemical composition of zinc alloys, and strongly increases with an increase of iron in zinc powder. This sensitivity to the presence of iron also varies with the composition of zinc alloys. We found that certain Al-Bi-In (ABI) lead-free zinc alloys display a zero-sensitivity to the presence of iron.

INTRODUCTION

In the 80's, a mercury reduction program was initiated by the zinc alkaline battery industry. This imposed significant changes in zinc anode formulation and alkaline cell technology (1, 2). Low gassing zinc alloy powders containing Pb, Bi, In, Al, Ga, and Ca were developed to meet these new requirements.

The shelf-life of alkaline cells is largely determined by zinc gassing caused by zinc corrosion in alkaline solutions. Zinc gassing was strongly inhibited in the presence of mercury, while mercury also improved interparticle conductivity in the gelled anode (3). In mercury-free technologies, it was shown that the unusual presence of some impurities strongly affects gassing of fresh zinc anode while partial anodic discharge (PD) increases zinc gassing. Under these conditions, more hydrogen gas is produced than in non-discharged anodes. PD gassing was found to strongly depend on the zinc alloy composition and the iron content (4, 5).

This work is devoted to the after-discharge (PD) gassing behaviour of zinc and the corresponding iron effect, and zinc alloying effects on out-of-cell after-discharge gassing of lead-zinc and lead-free zinc powders.

EXPERIMENTAL

Battery grade zinc powders were made by gas atomization of 99.997% zinc at the pilot plant located in the Noranda Technology Centre. Alloying elements (99.99%) were added to molten zinc. These powders were classified and characterized by the usual techniques, namely sieving analysis, chemical analysis (flame-AA), gassing tests, density measurements, and Scanning Electron Microscopy (SEM, Weiss apparatus).

Electrochemical measurements were performed in a three-electrode Plexiglass cell at room temperature with a Solartron 1286 potentiostat. The zinc anode, weighing about 1.6 g, in contact with a brass collector, was surrounded by a 100 mesh Nylon separator. The counter-electrode was made from nickel, and the reference electrode was Hg/HgO/37 wt% KOH - 3 wt% ZnO. The zinc anode contained 68 wt% zinc powder, 0.5 wt% Carbopol 940 - polyacrylic acid, BF Goodrich - and KOH-ZnO). A galvanostatic discharge was performed at C/8 or less. The gassing rate of partially-discharged zinc paste (at about C/18) was measured at 71°C by volume measurement or liquid displacement for 24 h. The anodic discharge of zinc was performed at an almost constant zincate concentration, since zinc was plated on the cathode while the zinc anode was anodically discharged. The depth of discharge (DOD) is expressed in percentage of the theoretical zinc anode capacity, i.e. 820 mAh g^{-1}.

Unless otherwise specified, the alkaline solution used was 37 wt% KOH containing 3 wt% ZnO, and the electrode potential is given with respect to Hg/HgO.

RESULTS AND DISCUSSION

Anodic discharge of pure Zn

The anodic discharge curve of pure gelled zinc at C/8 may be divided into three parts: initial, middle and final (Fig. 1). In the initial phase the potential displays a strong time dependence, which is characterized by a linear $E\text{-}t^{1/2}$ behaviour up to around 10% depth of discharge. This time dependence was also reported for several zinc alloys or amalgamated zinc anodes (5) or real alkaline cells (6). The same behaviour was also observed during large-scale anodic discharge of zinc paste.

Wen et al. have demonstrated (7) that this $E\text{-}t^{1/2}$ behaviour is expected at the early stage of a galvanostatic discharge, if diffusion within the electrode sample is the rate-determining process. In other words, a steady-state is established at the electrode-electrolyte interface by exchange of reactants between this layer and the electrolyte, and diffusion occurs between the bulk electrode and the first layer. Wen's equation was

[1] $$(dE/dt^{1/2}) = [(2\,I\,Vm) / (Za\,F\,S\,(D\,\pi)^{1/2})]\,(dE/dC)$$

where I is the current by surface area of S, Vm is the molar volume of zinc electrode sample and dE/dC is the slope of the emf-composition curve. Assuming that Vm is 30 cm^3 $mole^{-1}$, and S is roughly 50 cm^2 g^{-1}, thus the chemical diffusion coefficient, D, is about 10^{-8} cm^2 s, as $(dE/dt^{1/2})$ is typically ranged from 1.0 to 1.5 for gelled zinc anodes discharged at 100 mA g^{-1} (Table 1).

As anticipated from eq[1], the $E\text{-}t^{1/2}$ slope was found to increase linearly with an increasing discharge rate (5). On the other hand, the chemical composition of the zinc powder did not have any significant effect on this slope, probably because powder composition does not affect the surface area, the molar volume, or the diffusion coefficient. However, the absence of a gelling agent induced an increase of this slope, while the absence of zincate did not really affect this slope. The presence of a gelling agent may change the molar volume and the diffusion coefficient. The end of the initial phase would occur when zincate reaches a critical concentration, which is around 3 M zincate (8) or 10% depth of discharge for a gelled zinc having a density of 2 gZn cc^{-1}.

The polarization curves and gassing of Pb(500 ppm)-zinc powder were measured before and after anodic discharge. When the discharged paste is washed with water and acetone and dried at 50°C, zinc particles are found to have the same surface area, and almost the same uniform corrosion rate (I_{corr}) as fresh zinc particles (Table 2).

SEM images of discharged zinc powder in the absence of a gelling agent and ZnO in KOH are shown on Figures 2a and 2b, after the zinc powder was washed. The images reveal large channels and pores at the surface of zinc particles. The depth of these macropores increases with increasing DOD. SEM images (not shown) of discharged and washed gelled zinc (20% DOD) reveal that Carbopol particles cover much of the zinc surface and create bridges between particles. Macropores are also observed on the zinc surface. Light microscopy of the external layer of the discharged paste (in contact with an alkaline solution) shows that most of the attacked particles are large and irregular; the discharge state is also evidenced by the typical bluish colour in the external layer of the discharged paste.

Effect of anodic discharge on paste gassing

When the discharged paste of Pb(500 ppm)-Zn is washed with water and acetone and dried at 50°C, gassing rate of zinc particles seems to be lowered by the anodic discharge (Table 2). On the other hand, the measurement of the evolution of hydrogen from a discharged gelled zinc anode reveals that anodic discharge induces significant zinc anode gassing, which is largely amplified by the presence of iron (Fig. 3). This indicates that gassing is caused by some discharge products, which can be dissolved during aqueous washing. This product is likely to be ZnO.

The detrimental effect of zinc anodic discharge in alkaline solutions was first shown by Powers (9). More recently, Sato et al (10) discussed the effect of anodic discharge in terms of the electrocatalytic properties of solid zinc oxide for the hydrogen evolution reaction (HER). Moreover, Powers found that copper in zinc can enter the lattice of the oxide and enhance its electrocatalytic activity. However, the exact mechanism of ZnO formation by solid-state reaction or precipitation in concentrated alkaline solutions is still a matter of discussion.

At the present time, two mechanisms of PD gassing may be proposed: a) a catalytic activity of the ZnO precipitate or b) a catalytic activity of impurities accumulated and entrapped in a Zn/ZnO layer over the zinc. The mechanism should explain the combined effects of depth of discharge and iron in zinc. As a matter of fact, the catalytic activity of the ZnO layer cannot induce any iron effect on PD gassing. Impedance measurements performed during anodic discharge of zinc in concentrated alkaline solutions revealed that a porous oxide film always covers zinc metal during the anodic dissolution of zinc (11). This suggests that catalytic impurities may be accumulated and entrapped in a ZnO layer formed over zinc powder during the anodic discharge.

According to the model of impurity accumulation at the zinc interface, PD gassing should depend on both the concentration of critical impurities, such as iron and copper, and the depth of discharge. At the early stage of the discharge of spherical particles, the PD gassing of a 6 ppm Fe zinc powder at 5% depth of discharge should be similar to that of 2 ppm at 15%. This model is in a fairly good agreement with the data given in Table 3, where the same PD gassing almost corresponds to the same surface accumulation, as expressed by (DOD x Fe).

The proposed mechanism of impurity (or alloying element) accumulation at the zinc interface explains that excess PD gassing vanishes after discharged zinc powder is washed, and the accumulation layer is dissolved. On the other hand, this model does not exclude any catalytic contribution of ZnO.

Alloying effects on PD gassing

Table 4 shows that a small addition of Bi, In, Cd, and Mn to lead-zinc powder has negligible effects on PD gassing, while Al displayed some moderate inhibiting effects. Further additions of Bi and In increased PD gassing (Fig. 4). The inhibiting effect of aluminum was also evident in PABI (Pb-Al-Bi-In) and ABI (Al-In-Bi) powders (Table 5). Some specific combinations of ABI and PABI were found to be very effective in reducing PD gassing.

The effect of alloying elements on PD gassing is quite different from the inhibiting effect of Pb, Bi and In and the activating effect of Al on fresh powder gassing. This difference may be attributed to the gassing mechanism and the distribution of alloying elements. A surface analysis of fresh zinc particles indeed showed that bismuth, lead and indium were located in the grain boundaries area, while aluminum distribution is likely more uniform as it was not segregated into grain boundaries. Moreover, a SEM examination of corroded zinc surfaces showed that corrosion had occurred in the grain boundaries area, while SEM images of discharged zinc showed preferential dissolution of the inside of zinc grains, with the formation of deep pores (Fig. 2). These observations suggest that Bi, In and Pb are good inhibitors of gassing because they are located in the corrosion area and are not electrocatalytic for the hydrogen evolution reaction. On the other hand, these elements are not located in the zinc dissolution area, where aluminum is present and may interact with iron.

The local accumulation of additives, such as Bi and In, in the first zinc layer of discharged zinc particles should also be considered as a possible cause of the detrimental effect of these elements on PD gassing. Indeed, an excess of inhibitors may have detrimental effects on gassing. As aluminum should dissolve readily, such an accumulation should not occur with aluminum. Finally, it may be concluded that Bi, In and Pb have moderate inhibiting effects on PD gassing, while Al strongly inhibits PD gassing.

Alloying effect on the sensitivity of PD gassing to iron

Figure 5 displays the sensitivity of PD-gassing to the presence of iron in zinc powder. The effect is quite severe with Pb-Zn. There are then two ways to manage PD gassing:

a) reducing the iron content in zinc to less than 2 ppm, or
b) identifying synergic combinations of additives that exhibit lower sensitivity to iron.

Several ABI alloys were tested, at 250 ppm of lead and with no added lead. Figure 6 shows that Bi(200 or 500)-In(500)-Al (≥300) zinc alloys are relatively insensitive to the presence of iron, up to 30 ppm. The inhibiting effect of ABI systems is more significant when lead is also added, althouth the sensitivity of lead-free ABI alloys remain lower than that of Pb-Zn (Fig. 5). At low concentrations of iron, all the alloys display almost the same PD gassing. These curves also show that indium is not very efficient at 200 ppm. It may be concluded that 500 ppm In and 200-500 ppm Bi should be present, while more than several hundreds ppm of aluminum should be added to zinc in order to reach zero-sensitivity to the presence of iron. However, other factors, such as zinc anode performance under continuous or intermittent discharge, or shock sensitivity, should be considered before establishing the optimum composition of ABI zinc powder.

Conclusion

Gassing of partially-discharged mercury-free zinc powder in alkaline solutions was found to increase with increasing depth of discharge. This gassing rate depends on anodic discharge conditions and chemical composition of zinc alloys, and it strongly increases with increasing of iron in zinc powder. This sensitivity to the presence of iron also varies with the composition of zinc alloys. We found that certain Al-Bi-In (ABI) zinc alloys display a zero-sensitivity to the presence of iron. Aluminum was identified as a good PD-gassing inhibitor.

ACKNOWLEDGMENTS

The author thanks Richard Renaud for performing gassing measurements.

REFERENCES

1. D. von Borstel and D.H. Spahrbier. Prog. Batt. & Batt. Mat. **10** (1991) 38
2. H. Yoshzawa and A. Miura. Prog. Batt. & Batt. Mat. **12** (1993) 132
3. G. Dubé, R. Renaud, and J.Y. Huot. Prog. Batt. & Batt. Mat. **10** (1991) 151
4.T. Uemura et al. US Patent 5108494
5. J.Y Huot. Proc. Int. Power Sources Symp. (1993) 177
6. P.G. Cheeseman and M.G. Stock. Proc. Int. Power Sources Symp. (1984) 217
7. C.J. Wen, B.A. Boukamp, R.A. Huggins, and W. Weppner. J. Electrochemical. Soc. **126** (1979) 2258
8. F. Tye. in Electrochemical Power Sources. Primary and Secondary Batteries. M. Barak (Ed). IEE. Chap. 3
9. R.W. Powers. J. Electrochem. Soc. **118** (1971) 685
10. Y. Sato, M. Takahashi, H. Asakura, T. Yoshida, K. Tada and K. Kobayakawa. J. Power Sour ces **38** (1992) 317
11. C. Cachet, B. Saidani and R. Wiart. J. Electrochem. Soc. **139** (1992) 644

Table I. Effect of discharge rate, Carbopol (gelling agent) and ZnO concentrations on anodic discharge ($E\text{-}t^{1/2}$ in mV $s^{-1/2}$) of zinc paste containing pure zinc powder.

Discharge rate (mA g^{-1})	Carbopol (wt%)	ZnO (wt%)	$E\text{-}t^{1/2}$ slope
91	0.5	3	1.15
45	0.5	3	0.67
23	0.5	3	0.17
45	0.3	3	0.70
45	0	3	0.83
45	0.5	1.5	0.65
45	0.5	0	0.62

Table II - Effect of anodic discharge of Pb(500 ppm)-zinc powder on some physical and electrochemical properties of powder

	Gassing (cc/g-day)		BET	Polarization curves		
%DOD	Powder [(1)]	Gel		Eoc (mV)	Ecorr (mV)	Icorr (mA/g)
0	0.090	0.0058	< 0.01	-1403	-1433	0.14
7.5	0.074		< 0.01	-1383	-1420	0.35
15	0.016	0.250	< 0.01	-1381	-1434	0.20

[(1)] Discharged, washed (water, acetone) and dried powder. Gassing of discharged powder (100%) or a mixture of 20% discharged with 80% fresh in 45 wt% KOH; expressed in cc/g-day.

Table III - Combined effect of DOD and iron on gelled Pb-Zn anode gassing

DOD (%)	Fe (ppm)	PD (microL/g-day)
none	15	10
5	6	40
7.5	2	25
9	6	110
15	2	40

Table IV - PD gassing of zinc alloys containing Pb, Bi, Al, Cd, In and Mn. The iron content is around 2 ppm.

Pb	Bi	Al	Cd	In	Mn	PD Gassing
500						40
500	30					50
500	75					70
500	400					140
500				75		190
500				500		260
500		200		500		60
1000	1800					300
500		350				200
1000	1800	450				150
500	1800		20			220
500	1800	300	20			90
700			20			100
700		300	20			110
500		300			250	60
500					300	60

Table V - PD gassing of Pb-Bi-In-Al and Bi-In-Al powders at about 2 ppm of iron.

Pb	Bi	In	Al	PD Gassing
500	0	0	0	40
500	0	0	300	50
500	0	500	0	260
500	0	500	300	60
500	500	0	0	140
500	500	0	300	60
500	500	500	0	210
500	500	500	300	20
250	200	200	600	50
250	500	200	300	30
250	200	500	300	30
250	500	500	600	30
< 30	200	200	300	40
< 30	500	200	600	60
< 30	200	500	600	70
< 30	500	500	300	60
< 30	0	0	0	> 500

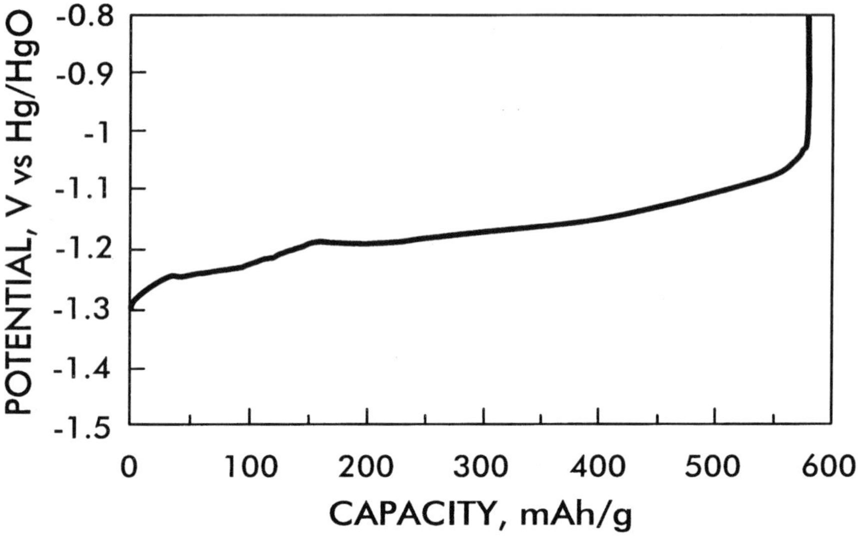

Figure 1 - Out-of-cell- discharge curve of Pb(500 ppm)-Zn gelled powder at 100 mA/g

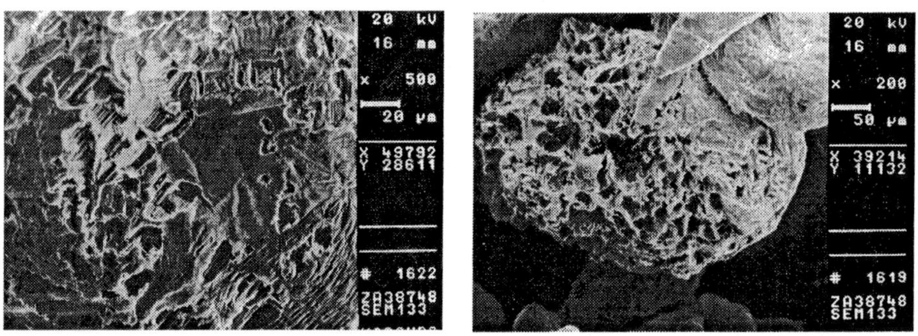

Figure 2 - SEM pictures of discharged zinc powder in the absence of gelling agent and ZnO in KOH at a) 1% DOD and b) 15% DOD after the zinc powder was washed.

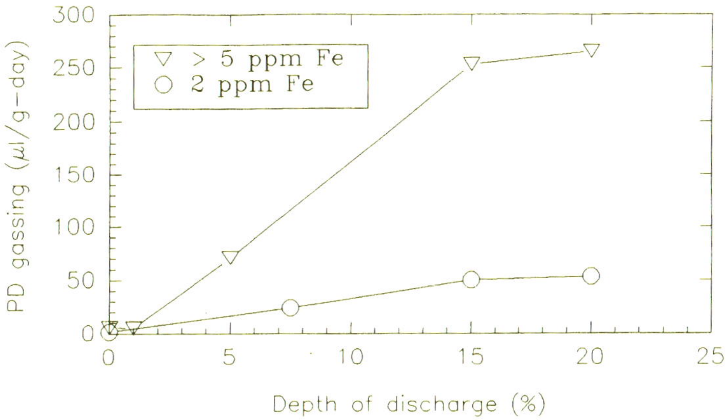

Figure 3 - After-discharge gassing rate of gelled zinc powder as a function of depth of discharge at 2 ppm and >5 ppm of iron.

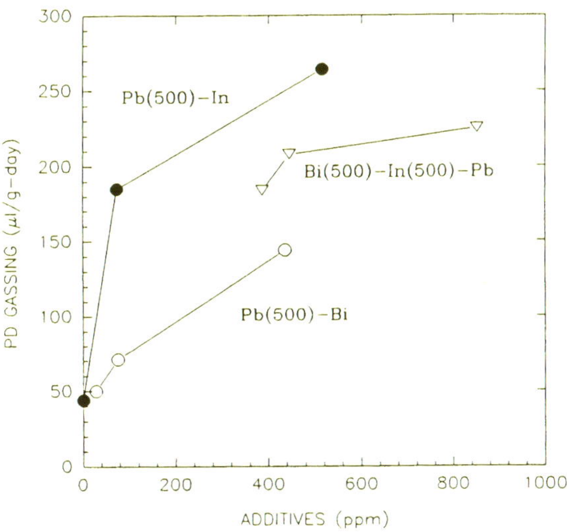

Figure 4 - Effects of additives concentrations on PD gassing of In-Pb, Bi-In-Pb and Bi-Pb zinc alloys.

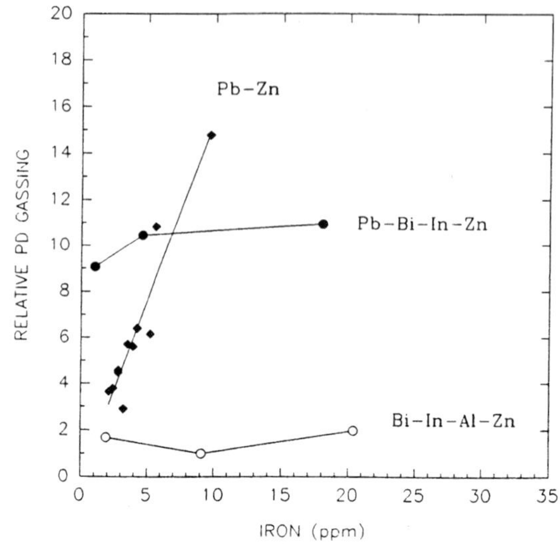

Figure 5 - Iron-effect on PD gassing of various types of zinc alloys

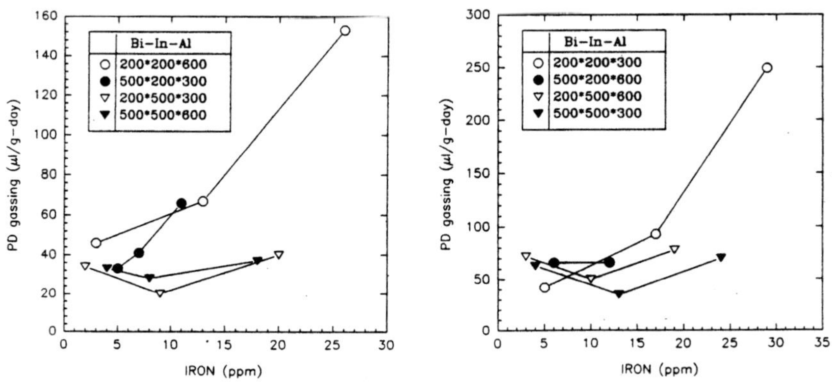

Figure 6 - Iron-effect on PD gassing of a) b(250 ppm)-Al-Bi-In, and b) ABI zinc powders.

Structure of Zincate Complexes in Concentrated Aqueous Hydroxide Solutions

K.I. Pandya (1), A.E. Russell (2), D. E. Sayers (1), J. McBreen (3), and W.E. O'Grady (4)

1. Physics Department, North Carolina State University, Raleigh, NC 27695, USA
2. Department of Chemistry, University of New Castle upon Tyne, New Castle upon Tyne, Ne1 7RU, UK
3. Department of Applied Science, Brookhaven National Laboratory, Upton, NY 11973-5000, USA
4. Code 6170, Naval Research Laboratory, Washington, DC 20375-5000 USA

X-ray absorption fine structure (XAFS) experiments were carried out to the structure of the Zn^{+2} species in concentrated aqueous solution of NaOH, KOH, RbOH and CsOH. The results show that the Zn^{+2} species are in tetrahedral configuration with a Zn-O bond distance of 1.97 A. No evidence of Zn-H_2O or Zn-Zn coordination shells is found. The importance of including the multiple scattering contributions to obtain correct EXAFS results is shown.

INTRODUCTION

The structure of Zn^{+2} complexes in alkaline electrolyte has been studied extensively using various electrochemical and spectroscopic techniques. Most of the structural studies show tetrahedrally coordinated $Zn(OH)_4$ species [1,2]. However, there continues to be a significant discussion about the structure because the solubility of zincate in concentrated hydroxide solutions formed by the anodic oxidation of zinc electrodes can be as much as three times the equilibrium solubility. In order to improve the performance of Zn-alkaline batteries it is necessary to study the nature of the dissolved Zn species. Various other species such as $[Zn(OH)_2(H_2O)_2]$, $[Zn(OH)_3(H_2O)]$, $[Zn(OH)_4 2H_2O]^{2-}$ and polymers formed by linking the $Zn(OH)_4$ tetrahedra via hydrogen bonding or bridging oxygens have been proposed. Most recently, it has been shown that a general formula for zincate ion is $[Zn(OH)_4 2H_2O]^{2-}$ with Zn^{2+} in tetragonally distorted octahedral coordination consisting of a planer $Zn(OH)_2^{2-}$ specie and two water molecules in the axial positions [3,4].

In the present work, we report XAFS studies of a series of concentrated aqueous hydroxide solutions saturated in Zn^{+2}. Our goal is to establish the structure of zincate ion and to demonstrated the necessity of including the multiple scattering processes in EXAFS data analysis.

EXPERIMENTAL

The XAFS experiments for electrolyte samples containing 0.74 M ZnO in 8.4 M NaOH, KOH, RbOH and CsOH were carried out in transmission mode on beamline X11A of the National Synchrotron Light Source at the Brookhaven National Laboratory, The XAFS experiments for a Zn/AgO cell containing supersaturated solutions in 12M KOH were carried out at LURE-CNRS.

RESULTS AND DISCUSSION

The radial structure functions (RSF's) for the 0.74M ZnO solutions in NaOH, KOH, RbOH and CsOH samples are shown in figure 1. In all cases the RSF's show a strong peak centered at ~ 1.6 A followed by a broad peak at ~ 3 A. Analysis of the first peak showed the Zn^{+2} has 4 oxygen neighbors at 1.96A. The second peak gives information about the long range structure of the Zn^{+2} species. The RSF's are identical for all the samples indicating that the local structure is not affected by the cations of the hydroxides. Figure 2 shows a Zn-O phase corrected RSF for ZnO in NaOH solution. The imaginary and the absolute parts of the RSF are coinciding for the first peak. However, for the second peak, they are out-of-phase by pi radians. Hence the possibility of second Zn-O shell due to waters of hydration is rejected. The possibility of a Zn-Zn shell is rejected because the EXAFS analysis yielded physically meaningless structural parameters. The origin of this peak is explained on the basis of multiple scattering processes occurring within the first shell for Zn in tetrahedral configuration [5]. Theoretical calculations carried out to model the multiple scattering processes are in good agreement with the experimental results. The EXAFS results for saturated and supersaturated solutions in KOH electrolyte were similar indicating that the local structure of zincate ions is same. However, small differences in the X-ray Absorption Near Edge Structure (XANES) were found indicating differences in the electronic structure of Zn.

CONCLUSIONS

The EXAFS results show that the Zn^2 ions are in tetrahedral configuration and the average Zn-O bond distance is 1.97 A. No evidence of bridging between tetrahedra was found. In order to determine the structure beyond the first coordination, EXAFS analysis must include the contribution of the multiple scattering processes.

ACKNOWLEDGMENTS

The authors gratefully acknowledge the support of the US Department of Energy, Division of Material Sciences, under the contract number DE-FG0589ER45384 for its roll in development and operation of beamline Z11A at the National Synchrotron Light Source (NSLS). The NSLS is supported by the Department of Energy, Division of Material Sciences under contract number AC02-76CH00016. This work was supported by the US Office of Naval Research.

REFERENCES

[1] T.P. Dirske, J. Electrochem. Soc., 101, 328 (1954).

[2] K.J. Cain, C.A. Melendres and V.A. Maroni, J, Electrochem Soc., 134 519 (1987)

[3] C. Debiemme-Chouvy and J. Vedel, J. Electrochem. Soc., 138, 2538 (1991)

[4] J. McBreen et al, Extended Abstracts Vol 89-2, Electrochemical Society, Pennington N.J. Page 12 Abstract No. 8 (1989)

[5] K.I. Pandya, Phys. Rev. B, 50 15509 (1994).

APPLICATION OF AQUEOUS SIMULATION SOFTWARE TO BATTERY RESEARCH

Andrzej Anderko, James W. Berthold, Marshall Rafal and Stephen J. Sanders
OLI Systems Inc., 108 American Road, Morris Plains, NJ 07950

The Corrosion Simulation Program (CSP) of OLI Systems has been applied to simulate the thermodynamic behavior of complex, highly concentrated aqueous solutions that are of interest in battery research. CSP is based on a comprehensive speciation of multiphase electrolyte systems coupled with accurate models for standard-state thermochemical properties and liquid-phase nonideality. In particular, solubilities of ZnO and zinc electrode rest potentials in the KOH-ZnO-H_2O system have been calculated and very good agreement with experimental data has been obtained. For a convenient analysis of the predicted results, the real-solution stability diagrams have been introduced. These diagrams show the stability areas of various species as functions of the potential and pH or concentrations of selected reactants. They combine the ease-of-use of the classical Pourbaix diagrams with rigorous thermodynamic calculations for nonideal multicomponent systems.

INTRODUCTION

Thermodynamic properties of electrolytes and electrochemical cells are of fundamental importance for the development of batteries. The basic properties of the KOH - ZnO - H_2O system, which is of primary interest for rechargeable zinc batteries, have been reviewed by Falk and Salkind (1) and Dyson et al. (2). There is continued interest in understanding the structure and thermodynamics of KOH + ZnO solutions in order to influence the solubility of ZnO (3, 4). For example, electrolytes of novel composition are studied to minimize the solubility of ZnO without compromising the cell power performance (5).

Therefore, there is a need for computational tools to simulate the behavior of complex, highly concentrated aqueous solutions. In particular, techniques are needed for modeling phase equilibria in systems with several possible phases and for predicting potentials in electrochemical cells. In this study, we examine the applicability of OLI Systems' Corrosion Simulation Program (CSP) for these purposes. Further, we establish models that accurately simulate the behavior of electrolytes that are of interest for the development of rechargeable zinc batteries.

THERMODYNAMIC MODEL

For the realistic modeling of multicomponent, multiphase electrolyte systems, a comprehensive thermodynamic model is of utmost importance. The model has to combine information about standard-state properties of all apecies that are present in the system with an accurate formulation for the excess Gibbs energy, which accounts for the solution nonideality. Additionally, it has to be coupled with an efficient algorithm for calculating phase equilibria in multicomponent systems that may contain an aqueous phase, any number of solid phases and, if necessary, a vapor and a nonaqueous liquid phase. These conditions are satisfied by the thermodynamic model that forms the basis of CSP. This model has been described in detail by Zemaitis et al. (6) and, more recently, by Rafal et al. (7). Here, we outline the essential elements of the model to explain how it can be applied to model the thermodynamic behavior of electrochemical cells.

The model considers all distinct chemical equilibria in the system of interest. In general, each of the equilibria can be expressed as

$$\sum_{i=1}^{n} \nu_i A_i = 0 \qquad [1]$$

where ν_i is the stoichiometric coefficient of species A_i and A_i is either a reactant or a product. The equilibrium state is reached when

$$\sum_{i=1}^{n} \nu_i \overline{G}_i = 0 \qquad [2]$$

where $\overline{G}_i$ is the partial molal Gibbs energy of the i-th species. Since molality (m_i) is convenient for measuring the concentration in electrolyte systems, $\overline{G}_i$ is related to concentration by

$$\overline{G}_i = \overline{G}_i^0 + RT \ln m_i \gamma_i \qquad [3]$$

where $\overline{G}_i^0$ is the standard-state partial Gibbs energy and γ_i is the activity coefficient. By combining equations [2] and [3], the equilibrium state of each reaction can be characterized by the equilibrium constant K, i.e.,

$$\ln K = -\frac{1}{RT} \sum_{i=1}^{n} \nu_i \overline{G}_i^o \qquad [4]$$

which is also related to the equilibrium concentrations by

$$K = \prod_{i=1}^{n} (m_i \gamma_i)^{\nu_i} \qquad [5]$$

This formalism can be applied to practical systems if the standard-state Gibbs energies of formation are available as functions of temperature and pressure (i.e.,

$\overline{G}_i^0(T,P)$) and the activity coefficients are known as functions of the composition vector **m** and temperature (i.e., $\gamma_i(\mathbf{m},T)$). From basic thermodynamics, the standard-state Gibbs energy of formation $\overline{G}_i^0(T,P)$ can be calculated if the following data are available:

(1) Gibbs energy of formation at a reference temperature and pressure (usually, T_r = 298.15 K and P_r = 1 bar);
(2) Enthalpy of formation at T_r and P_r;
(3) Entropy at T_r and P_r;
(4) Heat capacity as a function of temperature and pressure and
(5) Volume as a function of temperature and pressure.

Thus, the key to representing the standard-state properties over substantial temperature and pressure ranges is the accurate knowledge of the heat capacity and volume. A particularly useful equation of state for the partial molal heat capacities and volumes of aqueous species has been proposed by Helgeson and coworkers (8,9). It expresses the heat capacity and volume as functions of pure water properties and seven empirical parameters, which have been tabulated for large numbers of ions, complexes and neutral, both inorganic and organic, molecules (10, 11). The remaining thermodynamic properties are obtained by thermodynamic integration using the values of the Gibbs energy, enthalpy and entropy at reference temperature and pressure as integration constants. If the equation-of-state parameters are not available from the regression of experimental data, they can be estimated. For example, Shock and Helgeson (10) presented correlations for most solution species except for complexes. Sverjensky (12) developed an estimation method for complexes between metals and all univalent ligands except the hydroxide ion as well as selected divalent ligands. Another correlation exists for hydroxycomplexes. In addition to the seven equation-of-state parameters, these methods make it possible to predict the reference-state enthalpy and entropy if the reference-state Gibbs energy is known. These and other estimation techniques have been reviewed in detail by Rafal et al. (7).

The activity coefficient model used in CSP is an extended form of a model developed by Bromley (13, 14). The Bromley equation is a combination of the Debye-Huckel (15) term for long-range electrostatic interactions and an empirical expression for short-range interactions between cations and anions. In a multicomponent system, the activity coefficient of an ion i is given by

$$\log\gamma_i = \frac{-Az_i^2 I^{1/2}}{1+I^{1/2}} + \sum_j^{NO}\left[\frac{|z_i|+|z_j|}{2}\right]^2 \left[\frac{(0.06+0.6B_{ij})|z_i z_j|}{\left(1+\frac{1.5}{|z_i z_j|}I\right)^2} + B_{ij} + C_{ij}I + D_{ij}I^2\right] m_j \qquad [6]$$

where A is the Debye-Huckel coefficient which depends on temperature and solvent properties (6), z_i is the number of charges on ion i, I is the ionic strength (i.e., $I = 0.5\sum z_i^2 m_i$), NO is the number of ions with charges opposite to that of ion i, and B_{ij}, C_{ij} and D_{ij} are empirical, temperature-dependent cation-anion interaction parameters. Bromley's original formulation contains only one interaction parameter, B_{ij}, which is sufficient for systems with moderate ionic strength. For concentrated systems, the two additional coefficients C_{ij} and D_{ij} usually become necessary.

In the absence of experimental data, the B_{ij} parameter can be estimated from contributions of individual ions. In turn, these contributions can be predicted using a correlation with the entropy of ions. Further predictions can be made using a method due to Meissner (16), which makes it possible to extrapolate the activity coefficients to higher ionic strengths based on only a single, experimental or predicted, data point.

The Bromley model is restricted to interactions between cations and anions. For ion-molecule and molecule-molecule interactions, the well-known model of Pitzer (17) is convenient. Alternatively, the simple Setschenow (18) equation for the salting out effect can be used. However, these models are not accompanied by a technique for estimating missing parameters.

The combined OLI thermodynamic model is valid for temperatures up to 573 K, pressures up to 500 bar and compositions up to 30 m. For individual systems, the range of high accuracy may be limited due to a lack of experimental data for parameter regression (especially solution nonideality data) and, therefore, a greater dependence on the estimation of parameters. The CSP model is supported by an extensive data bank that covers the majority of existing inorganic species and a very substantial number of organic species, thus making CSP applicable to almost any battery chemistry of interest.

APPLICATION TO ZINC - ELECTROLYTE SYSTEMS

Solubility relations

Thermodynamic behavior of the ternary system H_2O - KOH - ZnO is of fundamental importance in zinc batteries. Solubilities in this system were experimentally investigated by Dirkse (19) and Dyson et al. (2). To predict the solubilities and other properties, the OLI simulation software is used for constructing a comprehensive model that defines the solution chemistry in this system. Table I lists all species in the H_2O - KOH - ZnO system and all independent equilibrium reactions that these species may undergo. As the equilibrium reactions, both chemical (e.g. hydrolysis) and physical (e.g. vaporization) transformations are taken into account. The reactions are arranged in such a way that they define the formation of more complex species from simpler species. Thus, the formation reactions are associated with all complex species.

Table I. Chemical species and independent reactions in the H_2O - KOH - ZnO system

Formula	Species type	Associated equilibrium reaction	
$H_2O_{(vap)}$	vapor	$H_2O_{(vap)}$	$= H_2O$
$H_2O_{(liq)}$	liquid	$H_2O_{(liq)}$	$= H^+ + OH^-$
$H_2O_{(s)}$	solid	$H_2O_{(s)}$	$= H_2O$
H^+	aqueous ion		
OH^-	aqueous ion		
K^+	aqueous ion		
$KOH_{(s)}$	solid	$KOH_{(s)}$	$= K^+ + OH^-$
$KOH \cdot H_2O$	solid	$KOH \cdot H_2O$	$= K^+ + OH^- + H_2O_{(liq)}$
$KOH \cdot 2H_2O$	solid	$KOH \cdot 2H_2O$	$= K^+ + OH^- + 2H_2O_{(liq)}$
Zn^{2+}	aqueous ion		
$ZnOH^+$	aqueous ion	$ZnOH^+$	$= Zn^{2+} + OH^-$
$Zn(OH)_{2(aq)}$	aqueous neutral	$Zn(OH)_{2(aq)}$	$= ZnOH^+ + OH^-$
$Zn(OH)_3^-$	aqueous ion	$Zn(OH)_3^-$	$= Zn(OH)_{2(aq)} + OH^-$
$Zn(OH)_4^{2-}$	aqueous ion	$Zn(OH)_4^{2-}$	$= Zn(OH)_3^- + OH^-$
$Zn(OH)_{2(s)}$	solid	$Zn(OH)_{2\,(s)}$	$= Zn^{2+} + 2OH^-$
$ZnO_{(s)}$	solid	$ZnO_{(s)} + H_2O$	$= Zn^{2+} + 2OH^-$

For battery applications, the H_2O - KOH - ZnO system is very concentrated with the ionic strength covering the full range of applicability of the CSP model (i.e., up to, or even exceeding, 30 mol/kg). Solubilities in such concentrated systems are very sensitive to activity coefficients. The most important binary interaction is between the K^+ and $Zn(OH)_4^{2-}$ ions because $Zn(OH)_4^{2-}$ is the predominant zinc species in highly alkaline solutions. Therefore, the B and C parameters of the extended Bromley model (cf. eq. [6]) have been regressed to reproduce the solubility of ZnO in KOH solutions for temperatures ranging from 243.15 to 317.75 K:

$$B_{K^+-Zn(OH)_4^{2-}} = 7.370 \cdot 10^{-2} - 4.752 \cdot 10^{-4} t + 7.989 \cdot 10^{-6} t^2 \qquad [7]$$

$$C_{K^+-Zn(OH)_4^{2-}} = -4.522 \cdot 10^{-4} - 1.207 \cdot 10^{-5} t + 1.04249 \cdot 10^{-7} t^2 \qquad [8]$$

In equations [7] and [8], t is the temperature in degrees C. The parameter D in eq. [6] was not necessary and was set equal to zero. Figure 1 shows that the solubility of ZnO in KOH is reproduced within experimental error. Additionally, Figure 1 shows the solubility curve for $KOH \cdot 2H_2O$, which also can precipitate in the H_2O - KOH - ZnO system. The solubility of $KOH \cdot 2H_2O$ was not taken into account in the parameter regression and, therefore, the calculated solubilities are slightly different from experimental data.

An important advantage of the aqueous simulation software is the possibility of screening the solubilities of solids in electrolytes with various chemical compositions. For example, studies have been reported on modifying the electrolyte composition in zinc batteries to reduce the solubility of ZnO (5, 20). This was achieved by replacing some of the KOH in the cell with various combinations of KF, K_2CO_3, K_3PO_4 and other salts. Our thermodynamic model makes it possible to predict the solubility of ZnO in mixed electrolyte solutions. This is illustrated in Figure 2 for the KOH+KF and KOH+K_2CO_3 solutions maintained at a total concentration of 10 mol/kg, but with varying proportions of the constituent electrolytes.

Rest potentials

Isaacson et al. (21) measured the rest potentials of the zinc electrode in concentrated KOH + ZnO solutions versus the Hg/HgO electrode. In view of the predominance of the $Zn(OH)_4^{2-}$ ion, the potential-determining reaction is

$$Zn + 4OH^- = Zn(OH)_4^{2-} + 2e^- \quad [9]$$

and the overall cell reaction is

$$Zn + 2OH^- + HgO + H_2O = Zn(OH)_4^{2-} + Hg \quad [10]$$

Thus, the cell potential can be predicted because the activities of all solution species (i.e., $Zn(OH)_4^{2-}$, OH^- and H_2O) can be easily obtained from equilibrium calculations as functions of the input amounts of KOH and ZnO:

$$E = \frac{\overline{G}^0_{Zn(OH)_4^{2-}} - 2\overline{G}^0_{OH^-} - \overline{G}^0_{HgO} - \overline{G}^0_{H_2O}}{2F} + \frac{RT}{2F}\ln\frac{a_{Zn(OH)_4^{2-}}}{a^2_{OH^-}a_{H_2O}} \quad [11]$$

The predicted potentials are shown in Figure 3 for various composition of the system. In Figure 3, the total molality of zinc in the solution ($m_{Zn,\,total}$) is almost, but not exactly equal to the molality of $Zn(OH)_4^{2-}$ because other zinc complexes are also present in small quantities (cf. Table 1). The agreement is very satisfactory and the average deviation from experimental data is 0.27%. The small deviations between the predicted potentials and experimental data may be due to the fact that ZnO occurs in different allotropic forms, which have slightly different thermochemical properties. Also, there can be a small inconsistency between the solubility data (from which the activity coefficient parameters were derived) and the rest potential data.

Real-solution stability diagrams

Traditionally, the Pourbaix diagrams (22) have provided a very useful tool for analyzing the thermodynamic behavior of electrochemical systems. However, the classical Pourbaix diagrams are subject to some inherent limitations. First, they do not recognize the solution nonideality and all equilibria are computed for arbitrarily assumed activities of solution species. Second, pH is introduced as an independent variable without regard to the chemical identity of the sources of the H^+ and OH^- ions, which determine pH. Third, the Pourbaix diagrams are limited to E_h and pH as independent variables whereas it is frequently worthwhile to examine the effects of other chemical species in the solution. These limitations are overcome by the *real-solution stability diagrams*, which can be produced with the help of our thermodynamic model. The real-solution stability diagrams show the predominance regions of various species as functions of the potential and either pH or molalities of solution species or input amounts of reactants. The compositions and activities are rigorously calculated using the OLI proprietary numerical solution algorithm which incorporates the aforementioned thermodynamic framework and data bank. The construction of such diagrams will be described in detail in a forthcoming paper. Here, we show their application to the Zn-ZnO-KOH-H_2O system.

Figure 4 shows a real-solution analog of the Pourbaix diagram for zinc, obtained by simulating the behavior of zinc in water with H_2SO_4 and KOH as pH-adjusting agents. The amounts of dissolved Zn species were assumed to be negligible and, therefore, their activities were set equal to 10^{-6}. The amounts of H_2SO_4 and KOH were varied to cover the whole pH range from 0 to 14. This diagram is very similar to the original Pourbaix diagram (22). On the other hand, Figure 5 shows how the stability ranges change if we add 1 mole of ZnO to 1 kg of H_2O and continue to use H_2SO_4 and KOH to vary the pH. In this case, the stability field of ZnO markedly expands and the field of $Zn(OH)_4^{2-}$ shifts to pH values greater than 14.

Figure 6 shows the behavior of a system containing 1 mole of ZnO in 1 kg of H_2O in strongly alkaline solutions. In contrast to Figures 4 and 5, the input amount of KOH is used instead of pH as an independent variable because it is more appropriate for highly concentrated solutions. As the input amount of KOH increases, the potential of metallic Zn declines with most of the decrease occurring for relatively small amounts of KOH (up to 3 moles). The vertical line between ZnO and $Zn(OH)_4^{2-}$ corresponds to the solubility of ZnO. Finally, Figure 7 shows the behavior of this system when the input amount of ZnO is varied for a fixed amount of KOH (i.e. 7.7 mol KOH in 1 kg H_2O). The equilibrium potential of metallic zinc is shown to increase almost linearily with the logarithm of the input amount of ZnO. Again, the vertical line between ZnO and $Zn(OH)_4^{2-}$ represents the solubility of ZnO in a 7.7 molal KOH solution.

CONCLUSIONS

As shown by the above examples, the CSP simulation system is capable of modeling practically important electrochemical systems with high accuracy. In particular, solubilities and equilibrium potentials can be accurately calculated for very concentrated solutions. This is made possible by the use of comprehensive models for standard-state properties and solution nonideality that are supported by extensive data banks of parameters. The results can be conveniently summarized in the form of the real-solution stability diagrams, which combine the ease-of-use of the classical Pourbaix diagrams with the real-solution behavior predicted by the OLI thermodynamic model. The application of CSP to the KOH - ZnO - H_2O system could be extended to almost any battery chemistry of interest due to the near-comprehensive coverage of various solution chemistries in the CSP thermodynamic model and in-place data bank.

REFERENCES

1. S. U. Falk and A. J. Salkind, Alkaline Storage Batteries, Wiley, New York (1969).
2. W.H. Dyson, L.A. Schreier, W.P. Scholette and A.J. Salkind, J. Electrochem. Soc., 115, 566 (1968)
3. F.R. McLarnon and E. J. Cairns, J. Electrochem. Soc., 138, 645 (1991).
4. J. McBreen, J. Power Sources, 51, 37 (1994).
5. T.C. Adler, F.R. McLarnon and E.J. Cairns, J. Electrochem. Soc., 140, 289 (1993).
6. J.F. Zemaitis, Jr., D.M. Clark, M. Rafal and N.C. Scrivner, Handbook of Aqueous Electrolyte Thermodynamics, American Institute of Chemical Engineers, New York (1986).
7. M. Rafal, J.W. Berthold, N.C. Scrivner and S.L. Grise, in Models for Thermodynamic and Phase Equilibria Calculations, S.I. Sandler, Editor, p. 601, M. Dekker, New York (1995).
8. H.C. Helgeson and D.H. Kirkham, Am. J. Sci., 274, 1089 (1974), ibid., 274, 1199 (1974), ibid., 276, 97 (1976).
9. H.C. Helgeson, D.H. Kirkham and G.C. Flowers, Am. J. Sci., 281, 1249 (1981).
10. E.L. Shock and H.C. Helgeson, Geochim. Cosmochim. Acta, 52, 2009 (1988), ibid., 54, 915 (1990).
11. E.L. Shock, H.C. Helgeson and D.A. Sverjensky, Geochim. Cosmochim. Acta, 53, 2157 (1989).
12. D.A. Sverjensky, in Reviews in Mineralogy, Vol. 17, I.S.E. Carmichael and H.P. Eugster, Editors, Mineralogical Society of America, Washigton, D.C. (1987).
13. L.A. Bromley, J. Chem. Thermodyn., 4, 669 (1972).
14. L.A. Bromley, AIChE J., 19, 313 (1973).
15. P. Debye and E. Huckel, Physik. Z., 24, 185 (1923), ibid., 24, 385 (1923).
16. H.P. Meissner, AIChE Symp. Ser. No. 173, 74, 124 (1978)
17. K.S. Pitzer, J. Phys. Chem., 77, 268 (1973).
18. J. Setschenow, Z. Phys. Chem., 4, 117 (1889).
19. T.P. Dirkse, J. Electrochem. Soc., 106, 154 (1959).
20. E.G. Gagnon, J. Electrochem. Soc., 133, 1989 (1986).
21. M.J. Isaacson, F.R. McLarnon and E.J. Cairns, J. Electrochem. Soc., 137, 2361 (1990).
22. M. Pourbaix, Atlas of Electrochemical Equilibria in Aqueous Solutions, Pergamon Press, New York (1966).

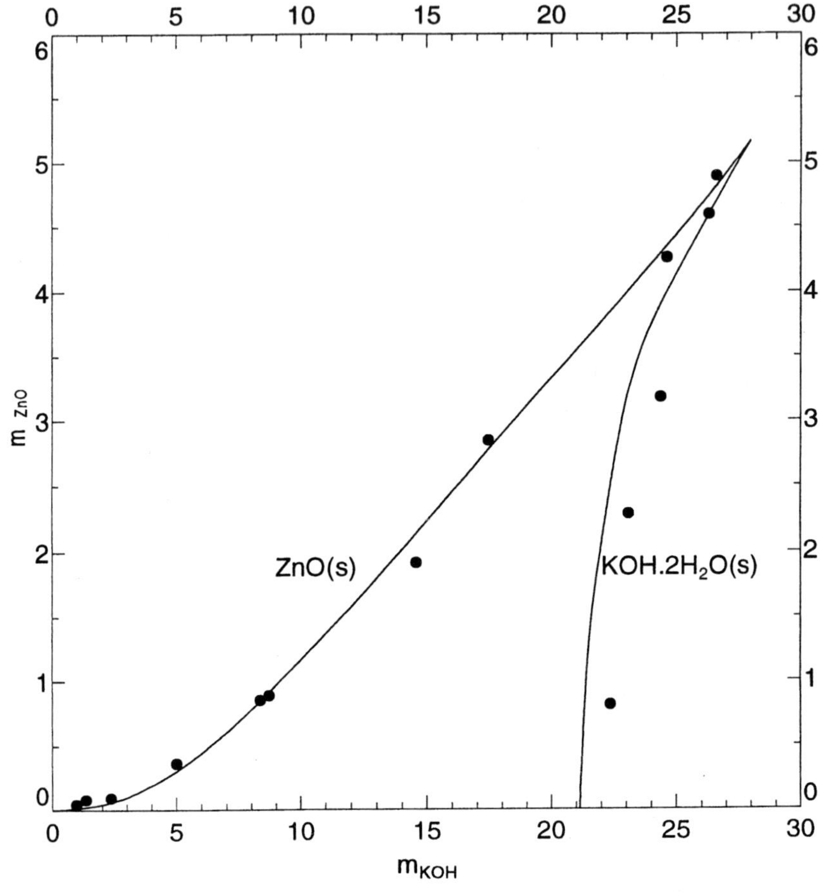

Figure 1. Solubility of ZnO and $KOH \cdot 2H_2O$ in the ternary system H_2O + KOH + ZnO at 298.15 K. The solid lines show the results obtained from the thermodynamic model and the circles denote the data of Dirkse (19).

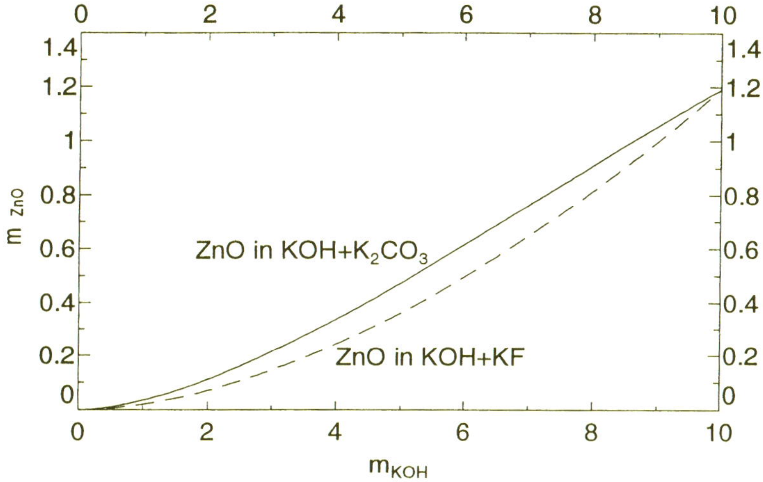

Figure 2. Solubility of ZnO in mixed electrolytes. The solid and dashed lines denote the solubility in $KOH+K_2CO_3$ and KOH+KF, respectively. The total molality of KOH and K_2CO_3 or KF is maintained at 10 mol/kg.

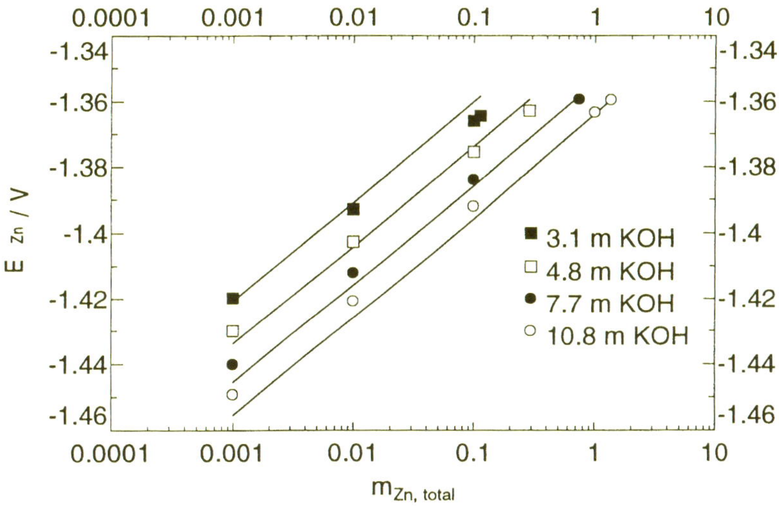

Figure 3. Zinc electrode rest potentials at 298.15 K vs. Hg/HgO predicted using the thermodynamic model (solid lines) and measured by Isaacson et al. (21, symbols).

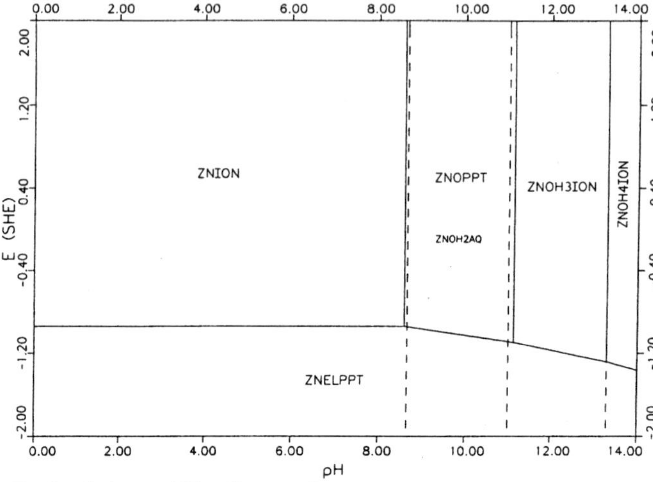

Figure 4. Real-solution stability diagram for Zn in H_2O with H_2SO_4 and KOH used as pH-adjusting agents. The activities of Zn species are assumed to be 10^{-6}. The symbols ZNION, ZNOH2AQ, ZNOH3ION, ZNOH4ION, ZNOPPT and ZNELPPT denote Zn^{2+}, $Zn(OH)_{2(aq)}$, $Zn(OH)_3^-$, $Zn(OH)_4^{2-}$, $ZnO_{(s)}$ and $Zn_{(s)}$, respectively.

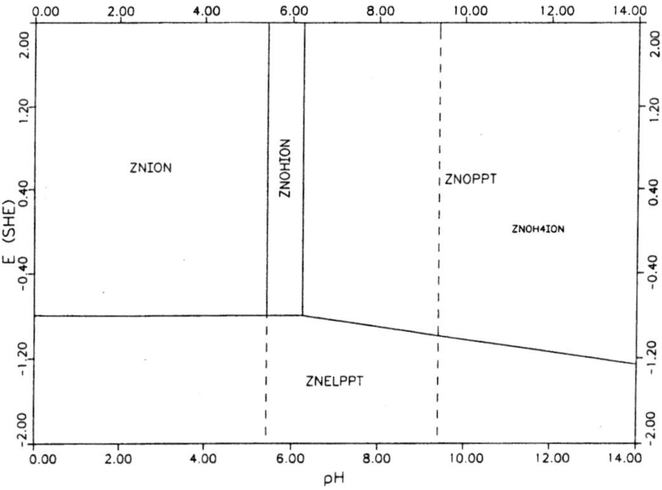

Figure 5. Real-solution stability diagram for a system containing 1 mol ZnO, 1 kg of H_2O and H_2SO_4 and KOH to adjust pH. The symbols ZNION, ZNOHION, ZNOH4ION, ZNOPPT and ZNELPPT denote Zn^{2+}, $Zn(OH)^+$, $Zn(OH)_4^{2-}$, $ZnO_{(s)}$ and $Zn_{(s)}$, respectively.

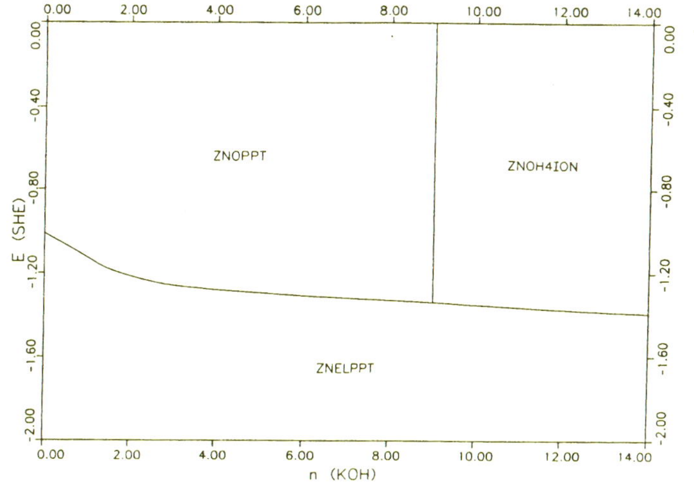

Figure 6. Real-solution stabilty diagram for a system containing 1 mol ZnO, 1 kg H_2O and varying amounts of KOH (from 0 to 14 mol). The symbols ZNOH4ION, ZNOPPT and ZNELPPT denote $Zn(OH)_4^{2-}$, $ZnO_{(s)}$ and $Zn_{(s)}$, respectively.

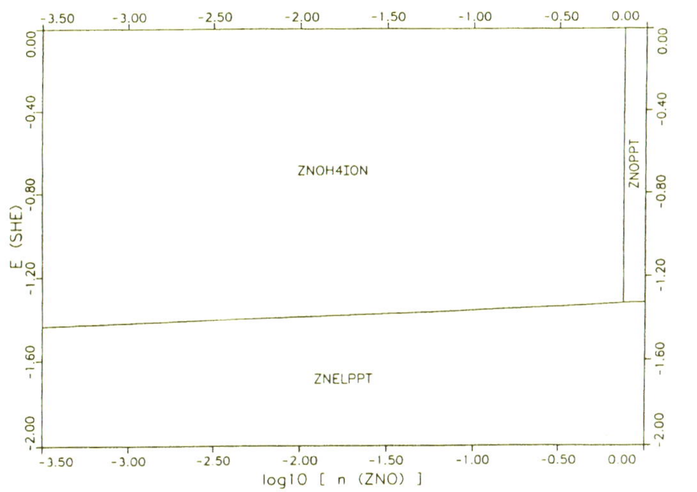

Figure 7. Real-solution stability diagram for a system containing 7.7 mol KOH, 1 kg H_2O and varying amounts of ZnO (from 0.00032 to 1 mol). The symbols ZNOH4ION, ZNOPPT and ZNELPPT denote $Zn(OH)_4^{2-}$, $ZnO_{(s)}$ and $Zn_{(s)}$, respectively.

CORROSION STUDIES OF ZINC ELECTRODE CURRENT COLLECTOR

L.Binder, M.Ghaemi, K.V.Kordesch
Institut für chemische Technologie Anorganischer Stoffe; TU Graz
Stremayrgasse 16, A-8010 Graz, AUSTRIA

The production of thin plate bipolar electrodes based on ABS foil filled with conductive carbon requires a good contact of the current collector towards the bipolar carrier foil and a good adhesion of the zinc electrode gel on the current collector. Electroless plating of the conductive foil on the anode side of a bipolar electrode can fulfill this request. The ABS foil can be plated with a thin copper layer after etching.
This basic copper layer must be thickened and covered with good conductive and corrosion resistant materials. Previously used materials such as tin must be
replaced by other materials due to its intensive gassing and high porosity. For that purpose some tin alloys were investigated in this work regarding their influences on corrosion and on the quality of the anode current collector. The tin alloys show better corrosion resistance and a more noble behaviour than tin.

INTRODUCTION

High discharge rates and fast charging reactions are expected in bipolar rechargeable cells. It was therefore interesting to investigate the corrosion effects and the electrochemical influence of tin alloys as anode current collector materials on electrode processes and the cycle live.

A first series of corrosion tests was carried out with mixed powders of pure metals and with binary alloy powders in contact with zinc. Additional to this, corrosion measurements of coated metal foils in KOH solution (10 N) were performed using a special pressure cell at room temperature and 45 °C. For comparison, measurements were carried out with zinc and tin powder, tin coated zinc foil and zinc coated tin metal sheet of the same geometry and size and applying the same test conditions.

Due to the lack of suitable bipolar prototypes the cited materials were finally tested as large surface current collectors (surface 12 cm^2) in a C-size battery fitted in the hollow space of a pressure cell equipped with a pressure sensor. The results of the current collector tests with tin alloys in C-size cells are compared with those of brass, tin, and electrolytically zinc coated brass foil. Cell pressure, number of cycles, cumulative capacity, and visual inspection of the anode current collector materials after use gave important informations concerning the suitability of binary and ternary tin alloys for battery use.

EXPERIMENTAL

Zinc corrosion in presence of powdered tin alloys

The required tin alloys were prepared by electrochemical deposition upon a conductive foil. The foil was not pre-treated for easier separation of the deposited metal layer from the substrate. The deposited alloys were powdered by a simple grinding operation and sieved to get a maximum particle size of 125 μm. For a single corrosion test 10 mg of alloy powder were mixed with 190 mg of pure zinc powder (Noranda, Type 215 200) by tumbling overnight. The mixture was introduced into a C-size battery container with sealed inner surface to prevent intermetallic contact of the powder mixture with steel. After adding 2 ml of 10 molar KOH the can was inserted in a pressure cell made of PMMA (polymethylmetacrylate) without anode current collector (Fig. 1). The pressure cell was equipped with a 250 mb-manometer and the tests were conducted at 45 °C (1). The amount of hydrogen produced by chemical corrosion could be calculated from the measured internal pressure and the known volume of the test cell and the reactants. The hydrogen volume was normalized to 1 g of zinc and divided by the elapsed time to estimate the corrosion rate. The same test was done with pure zinc powder, with zinc powder mixed with pure tin powder, and with a zinc powder/nickel powder mixture. The compositions of the test materials are given in Table 1 and the results of the corrosion tests are shown in Fig. 2.

Table 1. Composition of powders added to zinc for corrosion tests.

Metal type	conc. of tin [atom-%]	conc. of second metal [atom-%]
NiSn	55,4	44,6
ZnSn	76,5	23,5
CoSn	89,8	10,2
SbSn	63,6	36,4
Ni	----	100
Sn	100	----

The corrosion rate depends on time and different material compositions show different time functions. This can be seen in Fig. 3.

Corrosion of galvanically deposited layers

The following experiments can be divided into three sections: corrosion of zinc layers on tin or tin alloys, corrosion of tin or tin alloy layers on zinc foil, and corrosion of zinc on graphite foil. For the tests of section one (zinc on tin and tin alloys) zinc layers were deposited on previously deposited tin or alloy layers, both on conductive (carbon-)plastic foil. First experiments with nickel plated foils as carrier material revealed a strong influence of the nickel and these foils had to be abandoned. To improve the backside contact of the foil for the metallization process this side of the foil was layered with graphite. The corrosion tests were performed in the same test cell as the powder experiments and the same test temperature (45 °C) was applied. The values of Fig. 4 were obtained for samples with 9 cm^2 of surface area (exposed to KOH).

The experiments of section two (tin and tin alloys on zinc) were done exactly the same way with one exception. Tin or tin alloys were directly deposited on zinc foil without the previously used conductive polymer foil. Pure zinc foil of the same size (9 cm^2 of exposed surface area) and geometry was used as a standard. The results are collected in Fig. 5.

The intensive gas evolution of zinc in contact with graphite foil (Fig. 6) seemed to be a consequence of the mostly metallic impurities of even "pure" graphite (Table 2).

Table 2. Typical analysis of graphite foil (source: SGL-carbon)

Element	Conc. in ppm	Element	Conc. in ppm
Fe	300	Al	200
Cr	15	Ba	5
Ni	5	Ca	100
Mn	10	K	100
Pb	5	Na	50
Cu	2	Mg	100
Mo	5	Si	1300
V	2	Ti	30
Co	<1	S	1000
Sn	2		

Corrosion and electrochemical behaviour of plated current collectors

To obtain informations concerning the influence of plating alloys (tin basis) on the corrosion of anode current collectors under battery conditions these alloys were applied as surface layers on large area brass strips (12 cm^2) in C-size cells. The cathode (manganese

dioxide/graphite/acetylene black-composite) and anode (zinc powder/gelling agent/KOH) compositions were standard (2). After assembling the test cells in the pressure housing a first discharge via a constant load of 1.2 Ω was performed down to a cut-off voltage of 0.9 V ("zero discharge"). The following discharges were done via 1.8 Ω.
The cells were charged at a temperature of 25 °C during the first four cycles and at 45 °C for the fifth (and final) cycle. Charging voltage was 2.0 V in general. Charging time was four hours for the first three cycles and 16 hours (overnight) for the following cycles. In Fig. 7 the rise of internal pressure vs. time and cycle number is shown for the three most interesting materials.

RESULTS AND DISCUSSION

The experiments using metal powder mixtures showed the significant increase of hydrogen evolution by nickel, nickel alloy (NiSn), and the antimony alloy SbSn (Fig. 2). Though the corrosion rates for nickel and NiSn are decreasing with time (Fig. 3) these materials were excluded from the following layer experiments. Only tin, ZnSn, and CoSn were regarded as useful to undergo the subsequent tests. These tests should be a next step on the way to practical application in future bipolar batteries. This is the reason for changing from powders to thin layers deposited on conductive (graphite doped) polymer foil. According to the projected use in anode current collectors zinc was plated on the previously deposited tin and tin alloys. Fig. 4 shows clearly that the tin/cobalt alloy promoted zinc corrosion despite the fact that powder experiments did not manifest such a reaction. Obviously the mechanical properties and not the chemical composition of the deposited film are responsible for this effect. Low hardness, numerous cracks, and insufficient adherence to the substructure were found to be the most important characteristics. The visual inspection of the samples after the corrosion test revealed separation of the tin/cobalt layer from the carrier.

Consequently, only tin and tin/zinc alloy were subject to the following examination. In this procedure tin and tin/zinc alloy were plated on pure zinc foil and the results of the corrosion reactions (hydrogen evolution) were compared with those of zinc strips of identical size and geometry (9 cm^2, rectangular). Fig. 5 presents the low gassing rate of the bimetallic couple zinc-tin/zinc alloy. Additionally recorded cyclic voltammograms prove enhanced zinc dissolution and redeposition at more positive potentials.

Many observations sustained the doubts in results obtained when zinc came directly in contact with graphite felt or compressed graphite foil. Indeed, corrosion of zinc deposited directly on grphite foil was so fast that the 250 mb manometer had to be replaced by a 2.5 bar instrument. Though the shape of the hydrogen volume vs. time function in Fig. 6 looks similar to that on Fig. 5 there is a factor of about 40 in the vertical axis. This effect was attributed to the metallic impurities of graphite but this is not really verified. Similar promotion of zinc corrosion by different carbon materials has been found in earlier experiments too (3).

Just before starting the final tests of alternative anode current collector materials in complete C-size batteries by cycling and pressure measurements a new ternary Cu/Sn/Zn-alloy

("Optalloy") was proving excellent results in preliminary tests. This was the reason for adopting this metal as a candidate for the conclusive cycling test. Fig. 7 presents the results of this test for three materials: brass (which is usually applied), tin, and Optalloy with evident superiority of the latter. Later experiments (which are not included in Fig. 7) gave similar results for tin/zinc-alloy and for chemically zinc plated brass.
Very good results could be equally obtained by applying brass covered with a plated layer of tin/zinc alloy. Formation of a ternary copper/tin/zinc alloy on the surface of the brass foil cannot be excluded (similar to Optalloy?). Additional to the low gassing rate of C-size cells with Optalloy as anode current collector these cells showed the best performance on cycling. This positive effect seems to be created by a very low adherence of zinc and zinc corrosion products on the current collector. When the test cells were disassembled after five or six cycles neither passivating layers nor corrosion sites could be detected on used Optalloy current collectors.

CONCLUSIONS

If a material for anode current collectors in bipolar rechargeable manganese dioxide/ zinc batteries has to be selected Optalloy seems to be the first choice. Even when it is applied in very thin layers this material shows sufficiently high hardness, good corrosion resistance, and an inert behaviour in general. To deposit Optalloy on a conductive polymer carrier this material should be previously copper coated by an electroless plating process.

For conventional batteries with brass anode current collectors a significant improvement is possible by plating the brass spring or nail with tin/zinc alloy probably resulting in a ternary copper/tin/zinc alloy forming at the interface.

REFERENCES

1. V.K.Nartey,L.Binder,K.V.Kordesch,J.of Power Sources,**52**,217-222(1994)
2. M.Ghaemi,Dissertation, TU Graz (1995).
3. L.Binder,K.Kordesch,J.Electroanal.Chem.,**180**,495-510(1984).

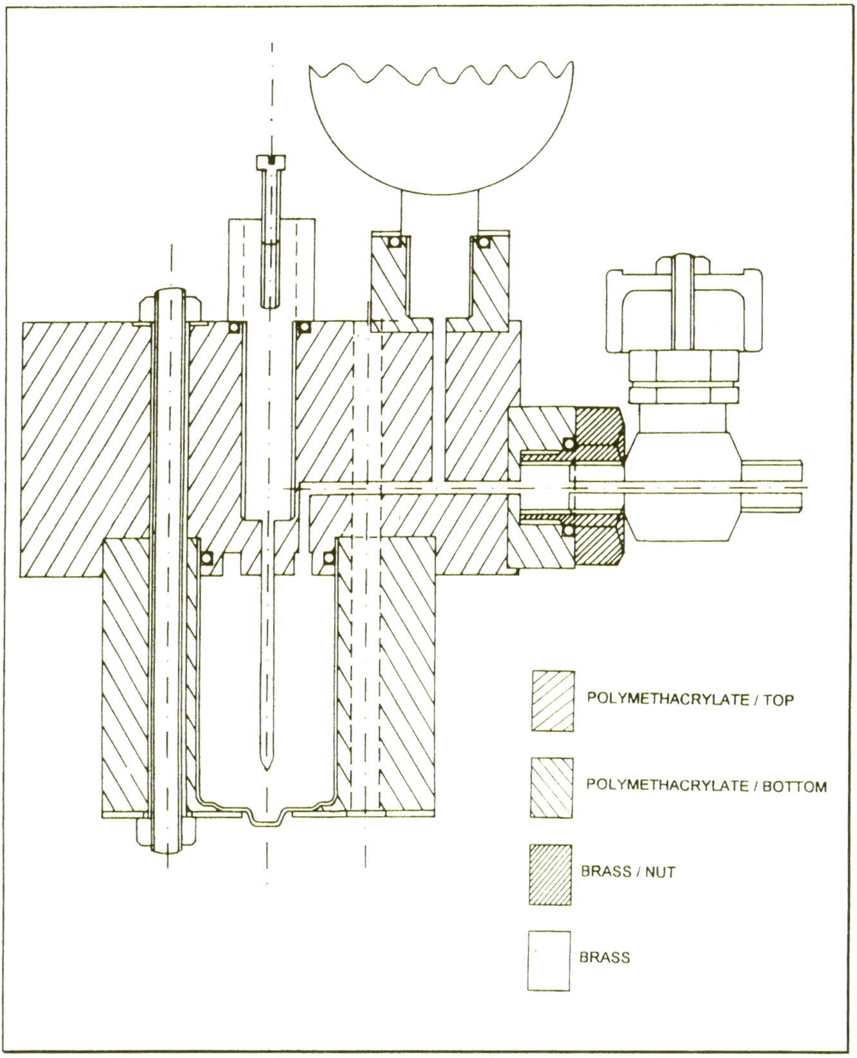

Abb. 1 Pressure cell for corrosion measurement.

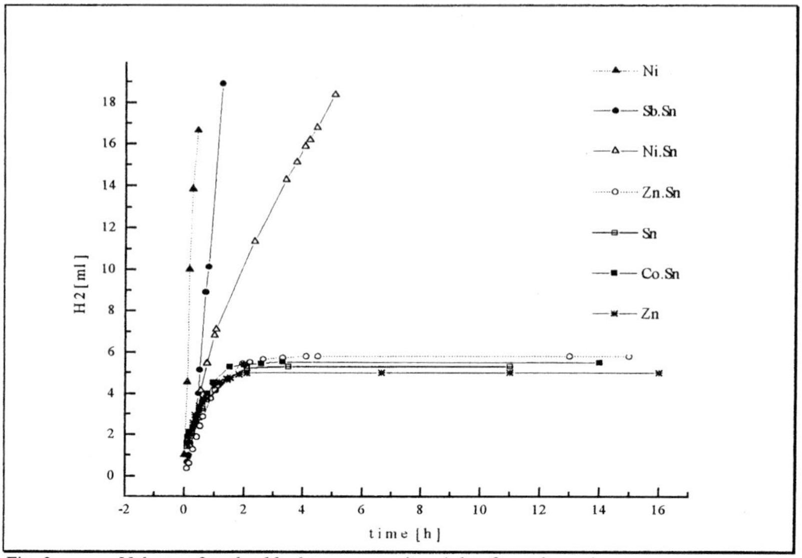

Fig. 2 Volume of evolved hydrogen per unit weight of metal powder.

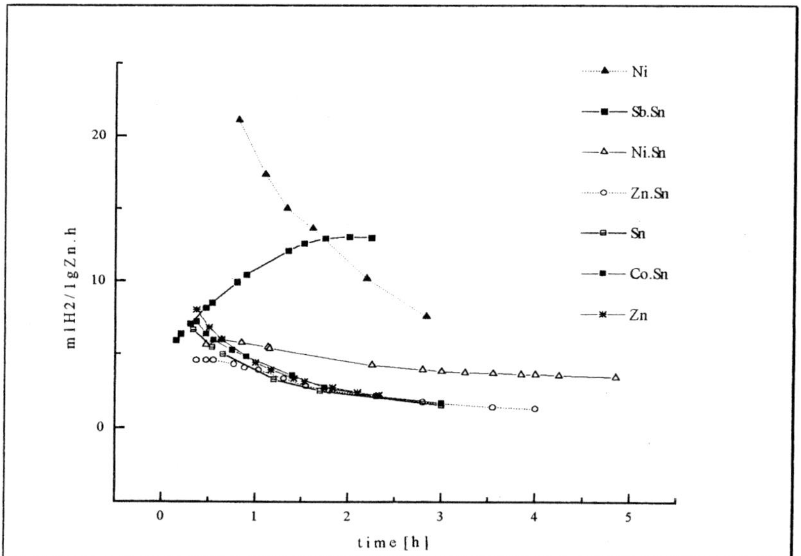

Fig. 3 Correlation of corrosion rate with time.

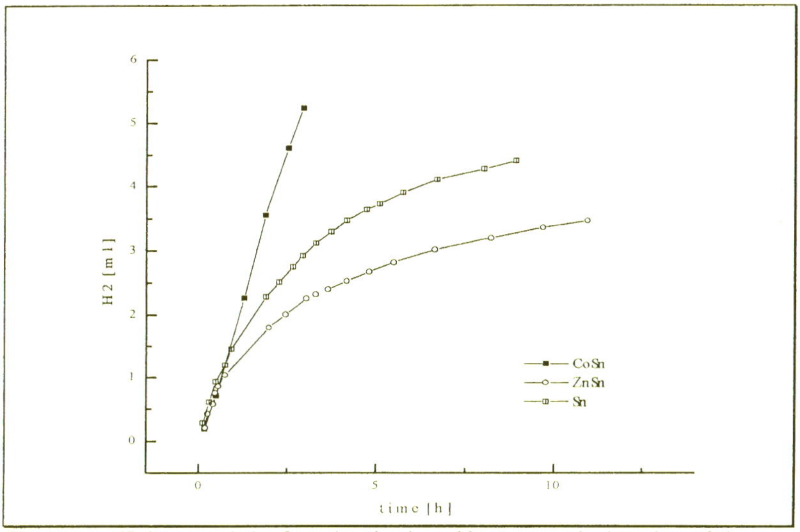

Fig. 4 Hydrogen evolution from zinc plated metal layers on conductive polymer foil (surface = 9 cm^2)

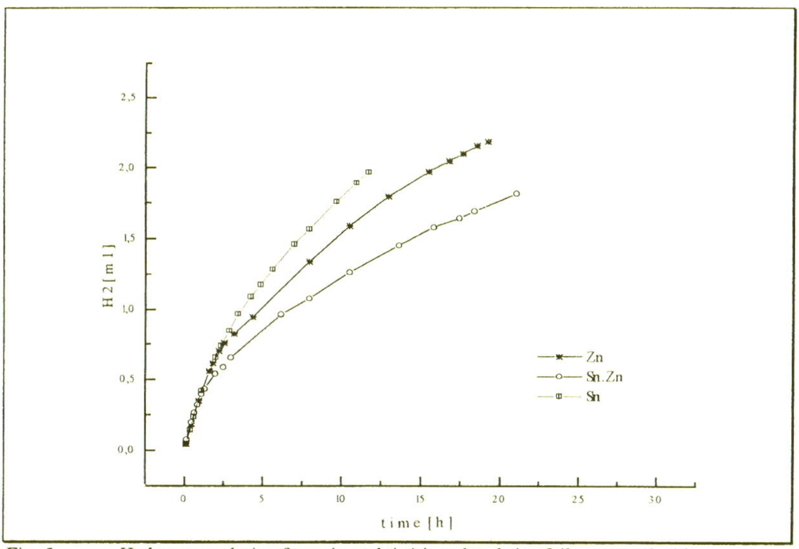

Fig. 5 Hydrogen evolution from tin and tin/zinc plated zinc foil compared with pure zinc (surface = 9 cm^2)

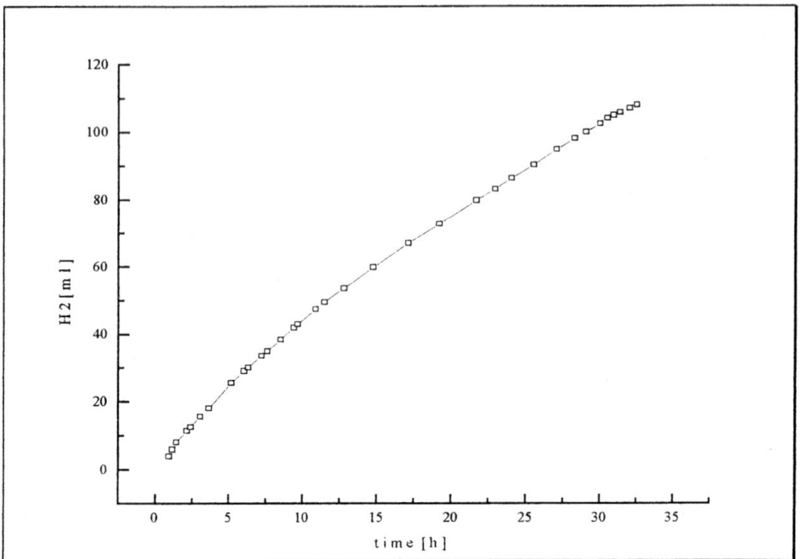

Fig. 6 Hydrogen evolution from zinc layer on graphite foil (surface = 9 cm^2)

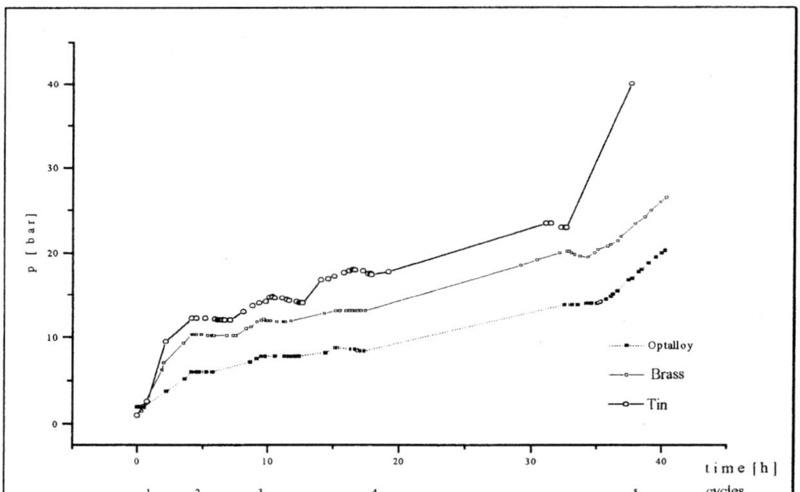

Fig. 7 Hydrogen evolution in a cycling C-size cell with different anode current collector materials.

PROBLEMS IN THE FIELD OF SILVER-ZINC STORAGE BATTERIES

V.S.Bagotzky, I.E.Yablokova, G.Z.Kazakevich*, T.A.Safronova
A .N.Frumkin Institute of Electrochemistry, Moscow
Industrial-scientific State Enterprise "Quant", Moscow

Some problems in the field of silver-zinc batteries are discussed: occurrence of voltage spikes during charging, decline in capacity when cycling at the lower voltage level, shape change of the zinc electrode, and influence of additives to the electrolyte solution on properties of the separator and on cycle life.

INTRODUCTION

Storage batteries on the basis of silver-zinc secondary cells are used for such applications, where high specific power and high specific energy performances of the battery, rather than its price, are of prime significance. In particular, they are used in some types of aircrafts and as buffer batteries in the power-supply system of some Russian sattelites and spacecrafts.

A detailed investigation of processes in silver-zinc batteries is important not only for the further improvement of these batteries, but also for the development of other types of storage batteries with zinc electrodes and alkaline electrolyte solution, as, *e.g.*, nickel-zinc batteries. Therefore the problems in the field of silver-zinc batteries can be divided into two groups: problems depending on the specific properties of the silver electrode, and problems common for all versions of rechargeable alkaline batteries with zinc electrodes.

VOLTAGE SPIKES IN CELLS WITH SILVER ELECTRODES

It is well known that the anodic oxidation of silver electrodes occurs at two distinct potential levels, which correspond to the formation of monovalent and of bivalent silver oxides (or, during cathodic polarization, to the reduction of these oxides). Therefore, alkaline storage cells with silver electrodes have well-pronounced two-step discharge and charge curves, and during operation the voltage varies over a wide range (changes up to 40%).

Another specific feature of silver electrodes, namely, occurrence (under certain conditions) of brief voltage spikes and dips, causes a further increase of this range. When charging, a small voltage spike always accompanies the transition from the lower step (Ag_2O formation) to the upper step (AgO formation). Higher values (hundreds of millivolts) of these spikes are observed, when the charge current is switched on within the transition region between the two potential steps or the initial

* – deceased

region of the second step, as well as after prolonged storage of charged electrodes. The duration of such spikes varies from a few to some tens of seconds.

These voltage maximums, which often occur during unplanned switching from discharge to charge or back, often adversely affect the operation of different electronic devices fed from the battery.

In the literature, different considerations were applied to the mechanism of extremum occurrence in the anodic polarization curves for silver in alkaline solution: an abrupt increase of electrode ohmic resistance due to the formation of a continuous Ag_2O layer [1], the formation of a "barrier layer" [2], an abrupt increase of true current density at small surface areas [3], the formation of an intermediate unstable oxide Ag_2O_3 [4], difficulties in the formation of the first nuclei for AgO crystallization in the Ag_2O crystal [5,6], a.o. No attempts have been made to reduce the amplitude and the duration of such transients.

Manifestation of Voltage Spikes

By experiments with smooth silver electrodes, and also with porous sintered silver electrodes (size 35x65x1.2 mm) in a solution of 10 M KOH it was shown:

(1) The voltage spikes arise on switching the current from discharge to charge in the transition region, as well as when after a certain storage time the current is switched on or rised abruptly in the region of the upper potential step. As a rule, the spikes after storage exceed the "transition" spikes by amplitude and duration (Fig.1).

(2) The spike's amplitude increases with an increase of charge current density up to 1-2 mA/cm^2, and tends to a limiting value with further current increase.

(3) The spike's amplitude increases with an increase in duration of the electrode storage in KOH solution (Fig.2).

(4) When charging at low temperatures (*e.g.* at 0° C), the amplitude and duration of the spikes sharply increase.

Influence of an Increase of Crystallization Nuclei Number.

In the present investigation it was attempted to reduce the voltage spikes by adding to the silver electrode oxides or metal salts with lattice parameters close to those of AgO. The best results were achieved with CuO and Bi_2O_3 [7]. By adding 5 wt.% of these oxides, the spikes could be reduced by a factor of 3-4. During charging at current densities up to 5 mA/cm^2, the cell voltage is limited by 2.0 V. However, Bi_2O_3 can not be used in practice, as its presence sharply increases the corrosion rate of the zinc electrode and the self-discharge of the cell.

The CuO additive were used in the production of silver-zinc batteries for sattelites with a life time of up to 6 months. After 6-7 months the positive effect of voltage tabilization disappears due to gradual leaching of CuO. Moreover, dissolved copper compounds deposit gradually on the zinc electrode and separator materials, leading to an increase of self-discharge and to a decline of cycle life.

Effect of current branching

Although CuO is isomorphous to AgO, in further studies it was found that the CuO action is not (not only) associated with the formation of crystallization nuclei for AgO. Voltage stabilization is also observed in cases, when CuO is not introduced into the electrode, but is placed into the system as a separate small electrode, connected in parallel with the silver electrode. It was shown that at the instant of a voltage spike the charge current is redistributed between the CuO and the Ag_2O electrodes. On the CuO electrode oxygen evolves at a potential, which is lower than the potential of further silver oxidation, and the current density on the silver electrode decreases abruptly. Within this time, a slow formation of AgO nuclei starts, and after a certain time the process is again completely "switched" to the silver electrode.

The Use of PbO Additive

In view of the foregoing results, the search of more chemically stable additives was continued. In so doing, not lattice parameters, but the ability of oxygen evolution in the potential range between the steady-state potentials of Ag_2O and AgO was used as criterion. The most encouraging results were obtained with the use of PbO additive.

Effect of lead additives on the behavior of silver electrodes was mentioned in several studies [8-10]. A decrease of electrode resistance and an increase of charge capacity were observed. After partial cathodic reduction of the electrode, three phases were found, namely, Ag_2O, PbO, and $Ag_5Pb_2O_6$. On evidence from [11], the latter compound has a low specific resistance of 3×10^{-3} Ω cm.

In this study it was shown that the addition of 2 wt.% of PbO results in an almost complete disappearance of the voltage spikes in the transition region and in the region of the upper voltage step (Fig.3). In contrast to copper additives, the effect is retained after prolonged cycling of the cells and/or after a prolonged storing of the charged cells at temperatures up to 50°C. The presence of lead compounds lowers also the voltage dip during discharge at high current densities, which is observed in the transition region. With respect to cycle life and overall capacity, batteries containing lead additives are approximately equal to standard batteries.

Measurements of the internal resistance of silver-zinc cells showed that in the transition region this resistance is much higher, than in the regions of the lower or upper potential steps. The addition of 5% CuO leads to a twofold reduction of the resistance, while with 2% PbO it is reduced by a factor of 5-8. Studies with a scanning electron microscope showed that electrodes with PbO additives have a finer grained structure than standard electrodes, both after a 1.5-month storage at +50°C and after cycling for 120 cycles at room temperature.

Measurements were also made with a supplementary electrode made of chemically prepared $Ag_5Pb_2O_6$ and connected in parallel with a silver electrode without additives. In this case no voltage spikes were observed neither. An almost 100% current branching to the supplementary electrode for a period of about 60 s was recorded.

Discussion of the experimental results on voltage spikes

From the above mentioned experiments it follows, that the voltage spikes on the silver electrode in the transition region arise from the lack of active nuclei for AgO crystallization. The spikes can be eliminated by an early introduction of the proper seeds serving as crystallization nuclei, or by a brief current decrease for a time sufficient for the slow formation of such nuclei. A temporary decrease of polarizing current is available by introducing CuO or PbO additives that divert the current to a "side" reaction of oxygen evolution in the transition region. It is also probable that the PbO additive exerts a supplementary influence, increasing the conductivity of the electrode and promoting an expansion of the reaction zone and a decrease in the true local current density.

CYCLING AT THE LOWER VOLTAGE LEVEL

Some electronic equipments in sattelites and spacecrafts require a high voltage stability. In some cases silver-zinc buffer batteries are used which are cycled only at the lower voltage level, corresponding to the formation and reduction of monovalent silver oxide. The lower capacity and energy density in such a cycling mode is compensated by comparatively low differences between the voltages during charging and discharging (the voltage deviations from the rated values are not higher, than $\pm 10\%$) and by a higher life time (a higher cumulative capacity).

For such kind of applications the main problem remains the problem of increasing the capacity at the lower potential level of the silver electrode. It is well known that during discharge of a completely charged silver electrode, the capacity at this level usually amounts to more than 80% of the overall capacity of the electrode. At the same time, during charging the length of the lower potential plateau is restricted. Depending on the current density used, the potential of the silver electrode rises to the upper level (corresponding to the formation of the bivalent silver oxide) as early, as after the oxidation of only 30-45% of the available silver to the monovalent oxide. During cycling the length of the lower voltage plateau further decreases. As during the first period of cycling the capacity of the battery is limited by the silver electrode, this capacity also declines. In Fig.4 the dependence of the battery capacity at the lower level (expressed in percentages as ratio to the theoretical capacity for the formation of the monovalent oxide) on the overall, cumulative capacity (expressed as $Ah \cdot cm^{-2}$) is shown. From curve 1 it can be seen that the capacity decreases at a rate of about $2\text{-}4\%/Ah \cdot cm^{-2}$.

One of the possible interpretations of this capacity decrease at the lower level can be connected with the gradual mass decrease of the silver electrode due to dissolution of silver oxides. However, experiments showed that this mass decrease amounts to $0.2\text{-}0.6\%/Ah \cdot cm^{-2}$, *i.e.* is much lower than the capacity decrease at the lower level. Furthermore, the rate of the overall capacity's decrease (during cycling at both voltage levels) is less, than the decrease at the lower level, and corresponds, approximately, to the losses of silver. Therefore, the capacity decrease at the lower level must be due to some changes of the properties of the silver electrode during cycling.

In order to increase and to stabilize the cycling capacity at the lower voltage level, the influence of different additives to the silver electrode and/or to the solution was investigated. The most pronounced effect was observed by adding silicates (about 5 g/l SiO_2) to the solution. In Fig. 4 curve 2 represent the corresponding capacity changes. It can be seen that from the very beginning the capacity is about twice as high, than without additives. Though this capacity also declines during cycling, it gives the practical possibility of a more prolonged cycling at the lower voltage level.

Silicates directly affect the properties of the silver electrode. In Fig.5 voltammetric curves for a smooth silver electrode in solutions of 10 M KOH without and with addition of 5 g/l SiO_2 are shown (sweep rate 100 mV/min). In presence of silicate a slight shift of the peak potentials can be observed, as well as a very sharp increase of the height of the first anodic maximum.

In Fig.6 porograms measured by the method of standard porosimetry [12] are presented for the initial silver electrode (curve 1) and for silver electrodes after 10 cycles in the alkaline solution without (curve 2) and with silicate addition (curve 3). It can be seen that after the first cycles (forming of the electrode) there is an appreaciable change in the porous structure of the electrode, *viz.* the formation of a high amount of pores in the radii range from 30 to 300 nm and a lowering of pore volume in the range 1-3 μ. In presence of silicate the amount of pores with lower size is appreciably higher, than in the pure alkaline solution.

During cycling no appreaciable further change of the shape of the porograms (of the ratio between pores of different size) is observed. The overall porosity at the first stage of cycling increases by about 10-20%. During the subsequent stages the porosity gradually decreases. This decrease is accompanied by a capacity decrease of the silver electrode during cycling at the lower voltage level.

At present electrolyte solutions with addition of silicate are widely used in Russia for the industrial production of silver-zinc batteries for space applications [13].

It is interesting to note that in presence of tellurates (1-50 g/l of tellur oxide TeO_2) the structure of the silver electrode changes in an even much more pronounced manner. During charging in such a solution the length of the lower voltage step increases considerably. On the voltammetric curve the anodic peak, connected with the formation of AgO, disappears almost completely (Fig.5). Unfortunately, this additive can not be used in silver-zinc batteries, as it leads to a sharp increase of zinc corrosion (and also for ecological reasons).

SHAPE CHANGE OF THE ZINC ELECTRODE

The phenomenon of "shape change" of the zinc electrode is well known to influence in a decisive manner the cycle life not only of silver-zinc batteries, but also of other alkaline batteries with zinc electrodes (*e.g.* nickel-zinc batteries). This phenomenon manifests itself as a deplacement of the electrode's active mass from one location to another, sometimes in upward, sometimes in downward or other directions. In the locations, where the active mass accumulates, its compression increases and the porosity diminishes. As a result, the utilization efficiency of the active mass decreases and premature passivation of the electrode is observed.

The data on shape change are very contradictory. This phenomenon is very sensitive to the cell design, particularily to the tightness of the electrode assembly and to the ratio between solution volume and amount of zinc. To obtain reliable results, statistical observations on a great number of electrodes and cells are needed.

In our experiments the degree of shape change was characterized by weight changes of three equal parts: at the top, the middle and the bottom of the electrode. For electrodes of size 38x99 mm it was shown that at current densities (for charge and discharge) below 5-7 mA/cm^2 the mass deplacement proceeds from the lower parts of the electrode to the higher parts. At higher current densities the deplacement occurs in the opposite direction. In addition, a deplacement from the inner parts of the electrode to peripherical parts, or *vice versa*, can be observed.

It is generally accepted that shape change is connected with the high solubility of zincate in the concentrated alkaline solution. The electrochemical reaction takes place not between metallic zinc and solid zinc oxide, but between zinc and dissolved zincate ions. The immediate cause of the mass deplacement is an uneven current distribution on the electrode's surface. Being different during charging and discharging, this leads to different rates of zinc dissolution and deposition at different locations. The influence of current distribution is confirmed by the observation of active mass deplacement not only within the limits of the electrode, but also between different electrodes of one and the same cell. In experiments with cells containing 24 electrodes, the initial scattering of the content of the active mass in different electrodes was ±4%. After cycling during 50 cycles the scattering increased to ±10%.

However, it must be noted that the reasons for such an uneven current distribution are up to now not clear. Undoubtedly, an initial uneveness of mass distribution after manufacturing of the electrode has an influence on this process, and during cycling this uneveness sharply increases. But the circumstance, that for all electrodes of a battery the main direction of deplacement is almost the same, points to the a prevailing influence of other factors.

Different methods for diminishing the influence of shape change were proposed. One of the oldest proposal was to add calcium oxide CaO or hydroxide $Ca(OH)_2$ to the active mass of the electrode. This additive diminishes the concentration of zincate ions in the alkaline solution and, hence, the degree of shape change. In the 50-ies in our country a small scale production of nickel-zinc batteries for mining lamps was organized, in which zinc electrodes containing CaO were used [14]. The use of this additive is also described in recent publications [15].

In our experiments with silver-zinc cells is was shown that, indeed, the addition of 5-15 wt.% CaO leads to a considerable lowering of shape change. At equal cumulative capacities the shape change in electrodes with this additive was up to six times lower, than in electrodes without additives. The directions of mass deplacement remained the same. Unfortunately, other properties of the electrode and the cell are also affected. At current density 20 mA/cm^2 the polarization of the zinc electrode in presence of CaO is by 50-70 mV higher, than without additives. At the same time the capacity and the cycle life decreases (2-4-fold decrease of the cumulative capacity). An appreciable increase in hydrogen evolution during charging was also observed.

These deteriorations of electrode performances are due to different reasons. A well pronounced increase of the resistivity of the active mass can be observed. But the addition of 3% graphite into the active mass in order to increase the conductivity resulted only in a slight decrease of the electrode's polarization (by about 10 mV), and did not affect the capacity decrease. In the active mass morphological changes take place: an increase of grain size, a decrease of surface area. But the main influence is probably the lowering of the zincate concentration and, hence, a lowering of the rate of the reaction path connected with dissolved zincate ions. It seems that a high zincate concentration is essential for high rate performances of the electrode and the battery.

Thus, not only the reasons for the phenomenon of shape change remain unclear, but up to now no reliable methods to overcome this difficulty are known.

FACTORS INFLUENCING CYCLE LIFE

Beyond the factors already discussed (gradual decrease of the efficiency of the silver electrode, shape change of the zinc electrode), the cycle life of silver-zinc batteries is in a decisive manner influenced by a complex of phenomena connected with the separator material.

In Russia a special version of cellophane (film n° 100) is used as separator material, which is produced by a proprietary method and whose properties differ from the properties of common cellophane used in most other countries. The swelling degree of this film in 10 M KOH is somewhat lower, than that of common cellophane, which leads to a slightly increased resistance (about 10-15%) of the film. At the same time the mechanical properties of this film in the alkaline solution are much higher, and the decline of these properties after prolonged storage (even at 50°C) much lower. The time of zinc dendrite penetration (measured by a special method) is about twofold higher. There is also a difference in the character of silver deposition in the films - silver ions are reduced and deposited mainly in the first layers (counting from the silver electrode), whereas the rate of silver penetration through the layers and deposition in the following layers is much lower. Most of the results described below were received in cells using this separator.

It was shown that the addition of 50-100 g/l of sodium tetraborate ($Na_2B_4O_7$) to the solution appreciable improves the properties of the separator and increases the cycle life of the cell (cf. curve 3 in Fig.4). This increase points to some "tanning" of the separator under the influence of tetraborate. This additive is now used for the production of most silver-zinc batteries in Russia [16].

It is interesting to note that the properties of the separator are also influenced by many other additives to the solution. So, the addition of 50 g/l of sodium tellurate Na_2TeO_3 leads to a much lower (about twofold) decrease of the mechanical properties of the film after prolonged storage in the alkaline solution. However, for reasons mentioned above, this additive can not be used in batteries.

Two kinds of cell failure after prolonged cycling and connected with the separator are usually considered: "hard" internal short circuits due to a zinc dendrite penetration through the separator with formation of electronic conducting "bridges" between the silver and the zinc electrodes, and "soft" internal short circuits due to a gradual metallization of the separator layers by silver deposition and the appearance of an electronic conductivity of these layers. Depending on the cycling conditions (current density, depth of discharge, etc.), either the first or the second influence prevail.

It was shown that there exists also a third possibility for internal short circuits of the "soft" type. During cycling, due to the formation of supersaturated zincate solutions and the diffusion of zincate ions through the separator layers, some amount of zinc oxide is deposited in the layers and between them. In contrast to chemically prepared zinc oxide, these deposits have a distinct (though low) electronic conductivity, which is probably due to some degree of doping with metallic zinc. This conductivity can also lead to losses of the capacity. It seems that in many cases "soft" selfdischarge with subsequent cell failure is due to the combined influence of metallization by silver of the first two separator layers and the accumulation of conductive zinc oxide in and between the other layers.

All these phenomena are very sensitive to the solution concentration. A lowering of the KOH concentration from 10 M to 7 M leads to a sharp decrease of the mechanical properties of the separator layer and to an increase of the metallization rate of the separator by silver. These factors, as well as a decrease of the efficiency of both the silver and the zinc electrodes, are the reasons for an appreciable cycle life decrease of silver-zinc batteries in the more diluted solution. .

REFERENCES

1. T.P.Dirkse, G.J.Werkema, J.Electrochem.Soc., **106**, 88 (1958).
2. A.A.Yakovleva, T.N.Borisova, and V.I.Veselovskii, Zh.Fiz.Khim., **36**, 1426 (1963).
3. B.D.Cahan, J.B.Ockerman, R.E.Amlie, and P.Ruetschi, J.Electrochem.Soc., **107**, 725 (1960).
4. A.Hickling and D.Taylor, Disc.Faraday Soc., **1**, 277 (1947).
5. P.Johns, H.Thirks, and W.Wynn-Jones, Trans.Far.Soc., **52**, 1003 (1956).
6. E.A.Mendzeritskii and V.S.Bagotzky, Dokl.Akad.Nauk SSSR, **142**, 127 (1962).
7. I.E.Yablokova and G.Z.Kazakevich, USSR Inventor's Certificate n° 194 156; (1966).
8. T.I.Dunaeva, F.I.Kukoz, and M.F.Skalozubov, in: Materyaly Konf. po Khim. Istochnikam Toka (Proc. Conf. on Chemical Power Sources), p.121, Novocherkassk (1966).
9. M.E.Alekseeva, V.N.Zgonnik, G.N.Fedorova, *et al.*, in: Sb. Rabot po Khim. Istochnikam Toka (Collected Works on Chemical Power Sources), n° 12, p. 58, Energiya Publ., Leningrad (1978).
10. L.I.Lyamina and N.I.Tarasova, Zh. Prikl. Khim., **10**, 2340 (1982).
11. A.Tvarusko, US Patent 3 650 832 (1972).
12. Yu.M.Volfkovich and V.S.Bagotzky, J.Power Sources, **48**, 327, 339 (1994).
13. V.P.Sirotinskaya, Yu.M.Volfkovich, V.E.Sosenkin, and I.E.Yablokova, Elektroteknicheskaya Promyshlennost', Ser.Khimich. i Fizich. Istichniki Toka, n° 6 /93, p.18 (1983).
14. N.A.Zhulidov and F.I.Yefremov, Vestnik Elektroprom., **2**, 74 (1963)
15. Jenn-Shing Chen, F.R.McLarnon, and E.J.Cairns, J.Power Sources, **39**, 333 (1992).
16. A.I.Shil'nikov, I.E.Yablokova, G.Z.Kazakevich, *et al.*, USSR Inventor's certificate n° 663 004 (1979).

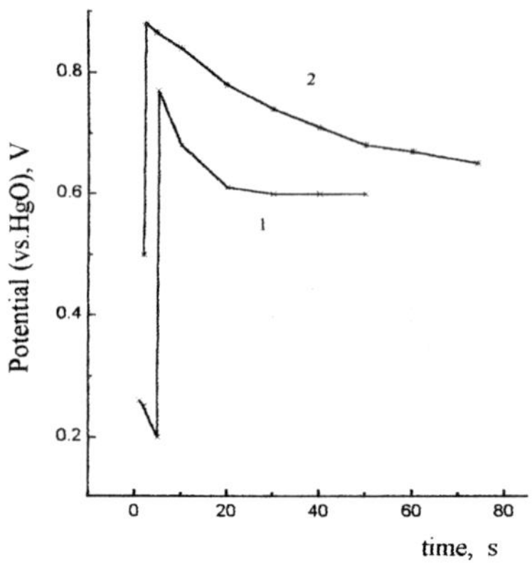

Fig. 1 - Potential spike for silver electrodes: 1 - in trans.region; 2 - after storage 40 h

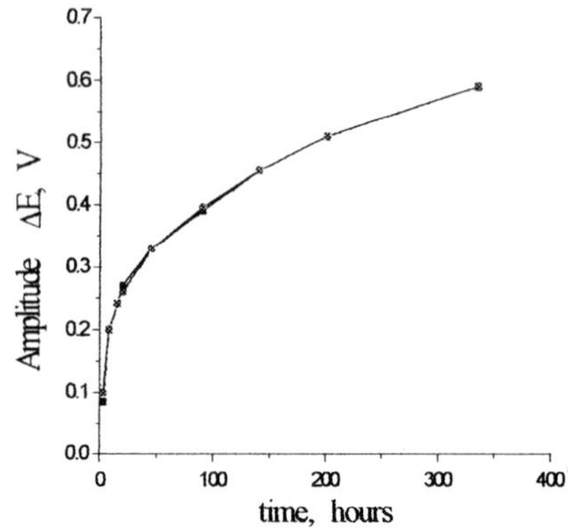

Fig. 2 - Dependence of spike amplitude on storage time; I = 3.5 mA/cm^2

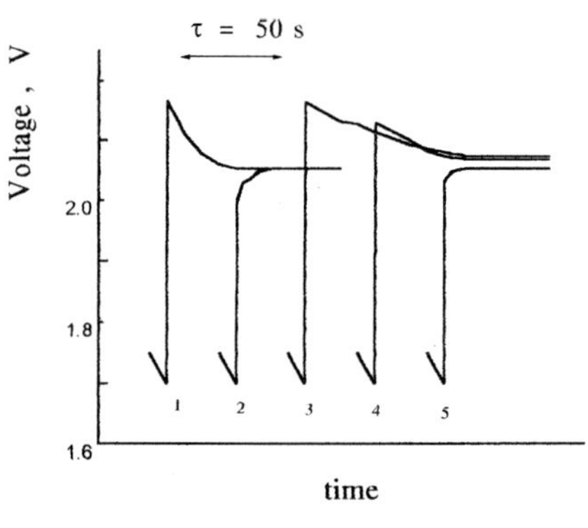

Fig. 3 - Voltage spikes at switching on anodic current 3.5 mA/cm^2: 1,2 - 40 cycles; 3-5 - 12 months storage; 1,2 - standard, 4 - with 5% CuO; 2,5 - with 2% PbO

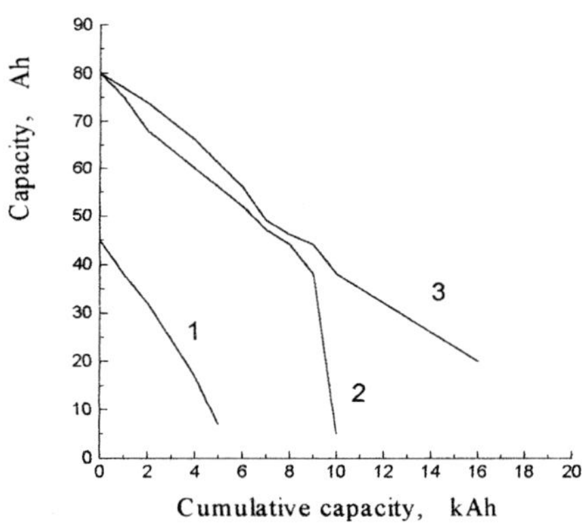

Fig. 4 - Capacity decline during cycling:
1 - without additives,
2 - with 5 g/l SiO_2,
3 - with 5 g/l SiO_2 + 50 g/l $Na_2B_4O_7$.

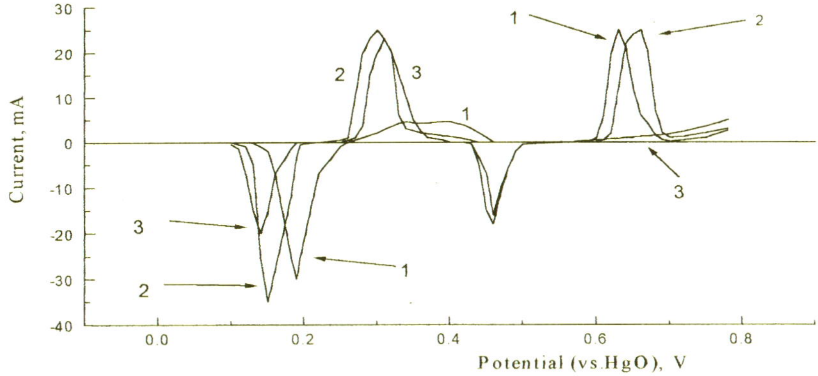

Fig. 5 - Voltammograms of silver electrodes in 10 M KOH:
1 - without additives; 2 - with 5 g/l SiO_2;
3 - with 5 g/l TeO_2 . v = 100 mV/min.

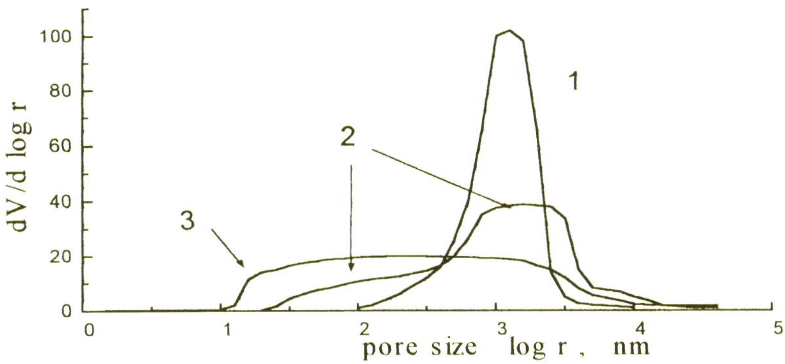

Fig. 6 - Differential porograms for silver electrodes:
1 - before cycling; 1,3 - after 10 cycles:
2 - without additives; 3 with 5 g/l SiO_2

PAST, PRESENT, AND FUTURE OF CELLULOSIC SEPARATOR IN SILVER-ZINC SECONDARY CELLS

Harlan L. Lewis
NAVSURFWARCENDIV Crane, Code 6094
300 Highway 361
Crane, IN 47522
Charles Grun, Alvin Salkind
Electrochemical Engineering Laboratory
Rutgers University
675 Hoes Lane
Piscataway, NJ 08854

The use of cellophane as a separator material in silver-zinc secondary cells goes back to 1941 and the work of the French chemist H. Andre. Although much work on this separator has occurred in the intervening years, there is a notable lack of specific evaluation studies on the importance of parameters such as residual degree of polymerization, crystallinity, etc., as they relate to cellulose source material and processing methods, and to the eventual cycle and wet-stand life of cells. Recently, chemists at the title centers have begun to evaluate basic properties of cellophanes as they relate to their physical parameters, to quantify the structural characteristics and apply that knowledge to cell cycle and wet-stand life by constructing a matrix of test cells with chosen separator films, for cycle evaluation. This paper reviews past literature work and present data on cellophane as a separator material, and projects future studies and their impact on silver-zinc performance.

INTRODUCTION

Pure cellulose (I) is a natural product of biosynthesis made up of D-(+)-glucose units connected by a glycoside linkage to C-4 of the next unit. It is found in wood pulps, cotton, and some grasses, e.g. Conversion of I to a modified form II (cellophane) by, e.g., the viscose process, is irreversible. Two structural families exist, parallel chain and antiparallel chain, where I is parallel and II is antiparallel. In both families, chains pack into planes in parallel fashion, but for I, successive planes are also parallel, whereas in II successive planes are antiparallel, as shown in Figure 1. The anti-parallel packing affords a more extensive 3-dimensional scheme of H-bonding so II is a more stable, lower energy structure, which partly explains why it cannot revert to I. Parallel chain structure can be retained in some transformations, e.g. formation of cellulose triacetate, (form III_I), and similar transformations occur for II ($II \rightarrow III_{II}$). But while $I \leftrightarrow III_I$ and $II \leftrightarrow III_{II}$, $II \not\leftrightarrow I$(1). Also, although the chain of D-glucose units is retained on going from $I \rightarrow II$, it is reported that the degree of polymerization (DP) is substantially reduced, by as much as 2-3X, as is degree of crystallinity, from 50% in wood pulp and 90% in cotton (I) to as low as 10% in II(1).

Other constituents in wood pulp-derived cellulose are hemicelluloses and lignin. Hemicelluloses are amorphous polymeric carbohydrates(2) made up of hexoses such as D-glucose, D-mannose, and D-galactose, and pentoses such as D-xylose and L-arabinose. Lignin is an amorphous phenolic polymer of three principal monomers, viz., coniferyl alcohol, sinapyl alcohol, and p-coumaryl alcohol(3). Wood pulps are 40-45% cellulose, 20-30% hemicellulose, and 20-30% lignin; cotton is ca. 95% cellulose(4,5).

Cellophane separator film from cellulose is typically used to wrap the plates in a rechargeable zinc/silver oxide cell. Its function is to provide a barrier to migration of $Ag(OH)_2^-$ ions while allowing free movement of the smaller electrolyte ions and water(6). It is also a sacrificial membrane because it is gradually hydrolyzed by the alkaline electrolyte and oxidized by the Ag(I) and Ag(II) species. To slow these processes, the following characteristics are desirable for the film: (a) a high degree of crystallinity to impart resistance to chemical attack and to reduce swelling in alkali, (b) a low degree of aldehyde character in the chains to inhibit oxidative degradation, (c) a high DP to retard onset of separator failure, and (d) the addition of Ag(0) to the film because it competes with the hydroxyl groups on the D-glucose units for reduction of bivalent silver, thus acting as a scavenger for Ag(II)(7).

HISTORY - PAST

Cellulosic Separator Technology

Since Andre first reported in 1941(8) and then patented(9) the use of a semipermeable membrane prepared from regenerated cellulose as a separator in silver-zinc rechargeable cells, many researchers have attempted to improve the properties of cellophane to a) reduce electrolyte and oxidative degradation, b) reduce silver migration to the anode, which leads to "soft" shorts, and c) inhibit zinc dendrite formation to prevent separator penetration with consequent formation of "hot" shorts. These efforts have taken two principal directions, viz., organic modification of cellulosic materials or cellophane and addition of inorganic modifiers to cellophane.

Organic modification. Many patents have been issued regarding incorporation of polyvinyl alcohol (PVA), often in combination with other organic substrates, as films bonded to cellophane through use of cross-linking agents or by radiative (uv or x-ray) processes, or incorporated directly into the cellophane film(10-15). Alternatively, cellulosic materials have been chemically modified with a variety of alcohols, amides, resins, etc., to improve wet strength and resistance to oxidative attack from silver (I) and (II)(16-30).

Inorganic modification. Salts in slurry form, with or without precipitation of elemental metals, have also been tried extensively(31-37).

Since current uses of cellophane in Navy applications are for the natural film without organic or inorganic modifiers, except for possible silver treatment, it appears that these efforts at improving cellophane properties were either largely fruitless or too expensive with respect to achievable benefits.

Cellulosic Separator Science

Over the past fifty years of use and attempted improvement of cellophane through the modifications referenced above, there have also been many studies of physical and chemical properties, to quantify the film properties as an assist to the modification procedures. Several books include sections on separator physical property measurements(38-40), and another(41) is devoted entirely to physical characterization of cellophane. Researchers have used a variety of techniques to characterize the properties of cellulosic materials and cellophanes with respect to strength and crystallinity, molecular (polymer chain) orientation, and degree of polymerization (DP). Among these have been:

1. Infrared and attenuated total reflectance spectroscopy to study extent of hydrogen bonding(42), glycerol content of films(43), and surface and subsurface molecular orientation resulting from various casting and drawing technologies(44,45).

2. Electrical properties to determine dielectric strength of films as a function of temperature(46), to study solute and solvent effects on conductance(47-49), and to obtain information on crystallinity via permittivity measurements(50).

3. DP and tensile strength (TS) measurements to determine degradation in KOH electrolyte(51), to compare cuprophane and viscose processing methods(52), to measure oxidative resistance after exposure to Ag_2O saturated electrolyte(53), to relate crystallinity and molecular orientation to DP and TS(54-58), and to relate acoustical property measurements to dynamic shear modulus and the Poisson coefficient(59). Ultrasound velocity has also been used to correlate molecular orientation, TS, and dynamic elastic anisotropy in cellulosic films(60).

4. Pore size and permeability studies of rate of capillary rise(61), for data on transport of silver oxides and decomposition rate for cellophanes(62), differential pulse polarography for zincate ion diffusion rates(63) and to characterize diffusion rates of various alkali metal hydroxides into cellophane films and cotton(64).

5. Crystallinity of cellulosic materials examined via an acid decomposition technique(65), x-ray and electron diffraction employed to study orientational distributions of crystalline phases(50, 66-68) and to measure glycerine and water content in cellophane casings(69), and orthogonal birefringence to obtain information on longitudinal molecular chain orientation during film swelling and stretching(70).

6. Electron microscopy to study surface properties as a function of machining technology(71-73), and to investigate the nature of the so-called "cuticle layer" at the surface of cellophane films(74,75). This cuticle layer has also been examined via a dye accessibility method(76).

7. Finally, thermogravimetry to evaluate E_{act} for pyrolysis in vacuo(77), and C-13 nmr to study the possible existence of supermolecular oxides in films(78).

PRESENT

Despite the history of testing and evaluation of cellophane films with the intent of improving alkaline rechargeable cell performance, over the past ten years a substantial decrease in cell life has occurred in cells built for Navy use. In past efforts to understand how the physical and chemical properties of various cellulosic materials contribute to the manner in which these materials behave in real silver-zinc rechargeable cells there has been an absence of coordination between measurement of separator properties and correlation of that data with cycle and wet-stand life data from a matrix of rechargeable cell tests, followed by thorough failure analysis. Thus, a program was initiated at the NAVSURFWARCENDIV Crane and Rutgers University, with financial support from a number of government entities, to study a set of properties for a variety of source-material and processing-method cellophanes, and then to build a matrix of test cells which incorporate the cellophanes for evaluation by cycle and wet-stand life testing, while assuring that statistically significant numbers of samples are used in all testing. The cell testing will be augmented by failure analysis on cells as their performance drops below capacity requirements and a redetermination of separator properties to determine cellophane failure processes.

Cellophane Property Testing

Tensile Strength for Dry and Alkaline-Soaked Films. The tensile strength of polymers is governed by the magnitude of the secondary valence forces operating between polymer chains which are reflections of the molecular structure of the polymer. So polarity of the mer unit, molecular symmetry, orientation of macromolecular chains (crystallinity), and length of polymer chains (DP) contribute to intermolecular forces and govern the ultimate strength of the separator material. Die-cut dog-bone shaped samples, 0.25 in across at the middle section by 6 in long, were tested "as is" or were soaked for 24 hr in 45% KOH (the standard electrolyte concentration), and then immediately placed in the test apparatus and the force necessary to break the film at the narrow section was obtained. The data in Table I are from eight samples per film, and the variation is 2X

TABLE I - Tensile Strength Data (Values in pounds)

Separator	Dry - MD*	Dry - TD*	45%KOH - MD	45%KOH - TD
Wood pulp A, clr	5.35±0.18	3.25±0.10	0.51±0.053	0.24±0.025
Wood pulp B, Ag trt'd (DuPont†)	5.04±0.26	3.10±0.16	0.81±0.075	0.33±0.046
Wood pulp C, Ag trt'd	5.44±0.48	3.41±0.05	0.63±0.11	0.30±0.055
Cotton E, Ag trt'd	6.01±0.14	3.29±0.19	2.01±0.12	1.21±0.075
Cotton F, clr	5.05±0.32	3.70±0.08	1.36±0.052	1.15±0.13
Wood pulp H, clr	4.97±0.10	3.54±0.09	0.40±0.02	0.24±0.006
Wood pulp I, clr	5.24±0.41	3.19±0.09	1.22±0.06	0.32±0.054

* MD = Machine direction, TD = Transverse direction

† This DuPont film is from an HR140 size cell manufactured in 1980, and is included for an historical perspective

the standard deviation of the mean (S). The dry film data are presented as baseline results, to compare to the KOH-soaked film data for relative changes in strength. In the wet-film data, the cotton-derived films (samples E & F) appear to retain the greatest strength, although one of the wood pulp films, sample I, is close in wet strength for the machine direction. It is also interesting that in comparison of samples A & C and F & E, which are similar except for the silver treatment, the treated films exhibit greater wet strength.

Swelling. Dimensional changes occur in separator films on wetting with electrolyte. In-plane changes should be as small as possible because they can lead to a shortening of life for rechargeable cells. Swelling allows increased large ion ($Ag(OH)_2^-$) migration due to increases in pore sizes. In Table II are given the data for thickness (swelling) after 24 hours soak in 45% KOH

TABLE II - Thickness (in mils) in 45% KOH

Separator	Thickness Dry	Thickness (24 hr soak - 45% KOH)		
		Low P	2 PSI	4 PSI
Wood pulp A, clr	1.04±0.02	2.52±0.08	2.41±0.07	2.36±0.12
Wood pulp C, Ag trt'd	1.1	2.86±0.12	2.81±0.08	2.76±0.06
Cotton E, Ag trt'd.	1.1	2.40±0.14	2.29±0.14	2.26±0.13
Cotton F, clr	1.1	2.28±0.14	2.15±0.10	2.13±0.10
Wood pulp H, clr	1.1	3.05±0.06	2.98±0.08	2.94±0.04
Wood pulp I, clr	1.0	2.83±0.09	2.81±0.08	2.76±0.06

starting from a dry thickness of 1 mil, at three pressure values from essentially 0 to 4 psi, which mimics plate pressure in an assembled cell. The data are for die-cut 2x2.5 in strips with the 2.5 in dimension parallel to the machine direction of the film, and are averages of eight samples. The cotton-derived films swelled the least. The 40% swelling difference for samples F and H, e.g., at 4 psi, would mean a substantial change in cell technology because cell volumes are typically limited in Navy applications, and in pore size because of the implications for large-ion diffusion.

X-Ray Diffraction (XRD). The degree of crystallinity(79) and the amount of included water(80) in the film are important to an understanding of film strength. The degree of crystallinity is substantially reduced by the viscose process; the greater the retained crystallinity, the stronger the film. The degree of water retention within a film is related to subsequent swelling in KOH, which affects porosity and is an indication of a film's effectiveness in preventing silver migration, because the greater the water retention, the less swelling will occur in KOH. In cellophane the degree of order determines the amount of swelling when an alkaline solution is absorbed(7), because interactions of OH^- and C-OH units aid in penetration of water into the structure, and $Ag(OH)_2^-$ absorption is also facilitated. Since the degree of swelling is also related to the degree of residual crystallinity, when the retained crystallinity is large, and retained water is also large, swelling is minimized and the primary penetration is by H_2O and OH^- (and K^+) as the lattice structure is maintained during separator life in a cell. In amorphous regions, the structure may be completely disrupted by water absorption so the polymer structure becomes a viscous solution with no distinct pore definition and penetration of large, destructive silver species is facilitated. Thus, large residual crystallinity and retained water in the structure are desirable as barriers to chemical attack by Ag(II) and Ag(I) ions, which serve to oxidize (degrade) the cellophane structure. As an assist to natural barrier properties, metallic silver may be deposited in the cellophane, either from redox involving -C-OH and/or -CHO, or coated onto the cellophane before use, to inhibit reaction between Ag(II) species and the film. As an alternative procedure, chemical crosslinking might also inhibit degradation as well as enhance crystallinity, but it could also inhibit that degree of swelling which makes alkali ion transfer efficient(7).

In the diffraction spectra of cellophane there are two principal 2θ regions of importance: one, at 2θ ≈ 12°, is a consequence of (101) plane interference, and a second, at 2θ ≈ 20 and 22°, is a $(10\bar{1})$ + (002) doublet (cf. Figure 2). Spectra for cotton and wood pulp cellulose are shown in Figure 3 (where the intensity differences are due only to sample size), a comparison of cellulose and cellophane spectra are shown in Figure 4, and cellophane and cotton cellulose diffraction patterns from the literature(79) are presented in Figure 5. The relative intensities for the two peaks listed are given in Table III. Now,(101) planes are preferentially oriented to the film sheet surface(80), and water

TABLE III - XRDS Data for Films and Celluloses

Sample	First Peak (12°)	Second Peak (22°)	Ratio R
Wood pulp A, clr	$0.89X10^3$	$0.41X10^3$	2.17
Wood pulp B, Ag trt'd (DuPont)	1.07	0.44	2.44
Wood pulp C, Ag trt'd	1.08	0.44	2.44
Cotton E, Ag trt'd	0.37	0.72	0.51
Cotton F, clr	0.58	0.81	0.72
Wood pulp H, clr	1.18	0.41	2.86
Wood pulp I, clr	0.84	0.45	1.89

is retained on the (101) surface of fibrils because most hydroxyl groups are located on the (101) planes. As a film dehydrates, uniplanar orientation of the (101) planes parallel to the film sheet surface occurs, and the value of $R = I_{(101)}/I_{[(10\bar{1}) + (002)]}$ increases. If the intensity ratio R is used as an indicator of retained water in a processed film, then the ratio for sample A corresponds to about 5% retained moisture while the cotton derived films (samples E & F) at the other extreme correspond to 25-30%(80). The data imply that films E and F have the greatest retained crystallinity and retained moisture (H-bonded to glucose -OH groups) so these two films should exhibit the greatest resistance to penetrtation by silver species, and consequently the greatest resistance to oxidative degradation. These results also conform to the measured swelling in Table II.

Degree of Polymerization (DP). The DP of cellophane is related to the film's ability to withstand degradation when used as a separator material in alkaline media because it is related to degree of retained crystallinity as discussed above. The tests were performed by intrinsic viscosity measurements in a cupriethylenediamine solution according to an ASTM method(81) or a method of Battista(82) using kinematic viscosity measurements. Two samples were determined for each material, and four measurements were made of each sample. Typically, the flow times varied by only 0.2-0.4 s in 250 s or more. The DP data are recorded in Table IV, and although the two methods gave significantly different results, the ordering is identical. They also coincide with the trends in the TS and XRD data, in that, for the cellophane films, the highest DP's were recorded for cotton film (sample F) and one wood pulp film (sample I), while the other wood pulp films gave substantially smaller values. Cotton and wood pulp cellulose data are presented as well, for comparison. Note also that the act of grinding cotton fiber to a powder at LN_2 temperature degrades the DP.

TABLE IV - Degree of Polymerization*

Sample	ASTM Calculated	Battista
Wood pulp A, clr	262	170
Wood pulp B, Ag trt'd (DuPont)	232	159
Wood pulp C, Ag trt'd	245	164
Cotton E, Ag trt'd	433	217
Cotton F, clr	437	219
Wood pulp H, clr	218	150
Wood pulp I, clr	462	239
Cotton fiber, powdered	706	339
Cotton fiber, as fiber	905	424
Wood pulp (Flexel)	418	230
Wood pulp (Viskase)	535	277

Differential Scanning Calorimetry (DSC). The enthalpy values associated with water loss and decomposition processes are related to crystallinity and the degree of polymerization of the films. The technique provides important information on retained water, structural order, and the residual lignin content which contributes to film weakness.

There are two principal phenomena exhibited in the DSC data in Table V. The first is

TABLE V - Differential Scanning Calorimetry Data

Separator	Dry Powder Samples		
	Energy		Rate
	0-200 C	Dec. range	Dec. range
Wood pulp A, clr	224 J/g*	7368 J/g	30.0 J/g/C*
Cotton F, clr	385	4806	17.2
Wood pulp H, clr	199	7302	29.7
Wood pulp I, clr	255	5249	21.3
Wood pulp B, Ag trt'd (DuPont)	216	7992	31.1
Wood pulp C, Ag trt'd	192	8593	34.4
Cotton E, Ag trt'd	324	6566	24.1
Cotton	181	5204	20.0

* J/g = Joules/gram, the total energy associated with the endotherm or exotherm and J/g/C = Joules/gram/degree Celsius, the rate of energy evolution for the decomposition process.

an endothermic process associated with loss of hydrogen-bonded water from the ambient condition films, for which the energy from 0-200C is given. The second is the decomposition exotherm, taken from 200-500C. The data for the silver-coated films are separated from the clear film data because of the possible influence of silver on the decomposition process.

The water loss energy values should by relateable to the XRD indications of residual water content. From the data for energy required (0-200C) for water loss for clear films, it does appear that the cotton film, sample F, which had the smallest R value in Table 3 contained the most moisture per gram of sample mass, and the wood pulp film, sample H, which had the largest R value, contained the least. Similar observations hold for the silver-treated films, between Tables III and V.

The decomposition energy values appear to reflect initial film purity, for the two films derived from cotton (samples E and F) exhibit the smallest energy terms, followed by the wood pulp sample I, which had a larger DP than the other wood pulp films. The larger energy values for the remaining clear films could be attributed to residual lignin content which, because of its aromatic nature, requires more energy for decomposition than polysaccharides. It is also curious that cotton has a slightly higher energy value for decomposition than the clear cotton cellophane film.

Thermogravimetric Weight Loss (TG). The data from the TG experiment are presented in Table VI, and again the silver-coated film data are separated because of the

TABLE VI - Thermogravimetric Weight Loss*

Separator	Dry Film Samples		
	Wt loss to 200C	Wt loss 200-525C	Resid wt
Wood pulp A, clr	11.64	72.77	27.23
Cotton F, clr	11.31	84.29	15.71
Wood pulp H, clr	10.07	73.50	26.50
Wood pulp I, clr	10.28	75.00	25.00
Wood pulp B, Ag trt'd (DuPont)	10.16	72.85	27.15
Wood pulp C, Ag trt'd	6.92	73.41	26.59
Cotton E, Ag trt'd	9.33	85.85	14.15
Cotton	4.64	90.40	9.60
Wood pulp, Flexel	6.54	87.65	12.35
Wood pulp, Viskase	5.01	84.16	15.84

* All weight losses are in percent. The weight loss from 200-525C is for the decomposition only, based on a dried sample weight, and residual weight is percent of original sample weight after moisture loss was taken into account.

residual weight of silver which impacts the residue weights. The initial weight losses for ambient condition film samples (to 200C) are inconclusive because they do not appear to reflect the trends noted previously for moisture content of the films. The starting temperature in the TG calance was typically 35C, and dry helium was blowing across the film samples during the weight stabilization period, which varied from one sample to the next, so the initial moisture content at the beginning of the heating process is not necessarily consistent with ambient water content. It is also noted that the weight losses to 200C for the raw materials are substantially less than for the films.

The weight loss from 200-500C (cf. Fig. 6) covers the decomposition region while the residual weight is the dry residue from the dry sample weight at 200C. In these data, the results for the wood-derived films are about the same while the cotton films have substantially lower residue weights, close to that for pure cotton. It is curious that for both types of cellophane films, the residue weights are larger than for the source cellulose materials.

Oxidative and Alkaline Degradation. Separator degradation in alkaline media occurs through alkaline hydrolysis, thermal depolymerization, and free radical oxidation. Degradation occurs by lowering the DP and probably also the degree of crystallinity, with consequent increase in amorphous character, resulting in turn in easier large ion diffusion and repetition of degradation processes. Degradation also results in formation of oxalate crystals which cause separator penetration, resulting in shorting, and general deterioration of the separator. These processes are relevant to both cycle life and wet-stand life for a cell. Measurements of degradation effects are performed by solution viscosity (DP) for alkaline processes and weight loss for oxidative processes(83).

The data obtained for oxidative degradation strongly suggested that the powder residues from the procedure with alkaline permanganate were slowly hydrolyzed by the acidic wash solutions, so that eventually the polymer residue was entirely converted to soluble glucose. The procedure needs to be reevaluated if it is to produce useful results.

The results for alkaline degradation, Table VII, indicate in a general way the loss of

TABLE VII - Alkaline Degradation Data for DP Changes

Separator	Initial	Four Hours		Twenty-four Hours	
	ASTM	N_2	O_2	N_2	O_2
Wood pulp A, clr	262	180	172	152	157
Wood pulp C, Ag trt'd	245	222	215	217	138
Cotton F, clr	437	311	165	167	76
Wood pulp H, clr	218	201	187	175	122
Wood pulp I, clr	462	386	357	292	223

DP with alkaline soak at temperatures of 70-80C. There are two observations available from the data: (a) the effects in the oxygen atmosphere are usually more pronounced than in the nitrogen atmosphere, and (b) the cotton-derived film degraded far more than the wood pulp films. This last result is surprising in view of the greater DP, TS, and crystallinity of the cotton film.

DISCUSSION

From the foregoing data and analyses it appears that if the film properties which have been reported to contribute to separator longevity in silver-zinc rechargeable cells, viz., KOH-wet tensile strength, swelling in KOH, degree of polymerization, and retained crystallinity and water content, are indeed the factors which contribute to long life for the separator film, then the sample ordering among clear films is:

$$E \approx I > A > H$$

and for silver-treated films:

$$F > B > C$$

The next step in the separator evaluation process is to build test cells of some small ampere-hour rating in a consistent manner from single batch silver and zinc plates with a selection of these separator films, to standard industry practices. A statistically significant number of cells for each separator type will be utilized, and the cells will be filled with 45% KOH and placed on a rigorous test regime where cycle life to 100% DoD and wet-stand life with periodic 100% DoD cycles are conducted. A standard procedure must be used, at constant temperature, so that the separators can be evaluated under cell life conditions.

Baseline properties will be established for dry, unformed cells and for formed cells (two formation cycles with electrolyte), and then as cells fail, failure analysis, silver diffusion, and separator properties such as TS and DP will be redetermined to establish degradation mechanisms under actual cell life conditions. From these data and the cycle and wet-stand life for the separator films, it should be possible to establish the connections between film properties and cell performance.

FUTURE

Some work on crossllinking of cellophane(84) and grafting of non-woven films to cellophane(85,86) has been reported recently in attempts to obtain better properties for cellophane, particularly to impede zinc dendrite growth. Work has also appeared regarding substitution of non-woven fabrics,(87,88) use of a combination of polyethylene and cellophane films,(89) and production and use of films from polysulfone/polyerher composites,(90) polyethylene/vinyl alcohol copolymers,(91) and a hydrocellulose/SL-080 graft copolymer film(92) as substitutes for cellophane. These initial efforts need to be extended and expanded to include other potential material combinations.

As they are developed, modified cellophanes and substitute films must be evaluated according to the same criteria which have already been used for the cellophanes evaluated previously, in order that physical and chemical properties can be compared. Then, those new films which show promise must be subjected to the same rigorous cell evaluation testing as is being used for the several different cellophanes cited above. Only in this way can the selection of separator films for optimum cycle and wet-stand life under both low- and high-rate conditions be advanced to the point where cells incorporating alkaline zinc chemistry will provide several hundred deep-discharge cycles with wet-stand lives of two-three years. These properties are the long-term goals of the development program.

REFERENCES

1. A. Sarko, in Cellulose - Structure, Modification, and Hydrolysis, R. A. Young and R. M. Rowell, Editors, p. 29, John Wiley & Sons, New York (1986).
2. M. Wayman, *ibid.*, p. 265.
3. K. Freudenberg, in Advances in Chemistry Series, No. 59, J. Marton, Editor, ACS, Washington, D.C. (1966).
4. R. A. Young, R. A Young and R. M. Rowell, *op. cit.*, p. 91.
5. E. Kiran, R. A Young and R. M. Rowell, *op. cit.*, p. 349.
6. H. P. Gregor, in Zinc-Silver oxide Batteries, A. Fleischer and J. J. Lander, Editors, p. 219, Electrochemical Society Series, John Wiley & Sons, New York (1971).
7. R. E. Post, *ibid.*, p. 263.
8. H. Andre, Bull. Soc. Franc. Electriens (6th Series), **1**, 132 (1941).
9. H. Andre, U.S. Pat. **2,317,711** (27 Apr. 1943).
10. M. N. Yardney and M. E. Kagan, U. S. Pat. **2,635,127** (14 Apr. 53).
11. H. Winkler, Ger. (East) Pat. **18,764** (3 May 1960).
12. M. Fukuda and T. Iwaki, Jap. Pat. **1824** (6 Apr. 1959).
13. S. Matsuno, Jap. Pat. **2767** (21 Apr. 1959).
14. M. Mendelsohn and F. Solomon, U. S. Pat. **3,013,100** (2 May 1957).
15. K. Sugimoto and K. Takeda, Ger. Offen **2,535,863 (26 Feb. 1976).**
16. R. S. Locke and R. J. Daly, U. S. Pat. **2,504,208** (18 Apr. 1950).
17. E. R. Dillehay, U. S. Pat. **2,531,504** (28 Nov. 1950).
18. E. Fairclough and C. C. Rose, U. S. Pat. **2,677,008** (27 Apr. 1954).
19. R. C. Shair, P. F. Bruins, and H. P. Gregor, Ind. Eng. Chem. **48**, 381 (1956).
20. T. Fukuda and S. Oda, Jap. Pat. **1470** (18 Mar 1959)
21. M. Mendelsohn and F. Solomon, U. S. Pat. **2,964,584** (13 Dec. 1960).
22. E. S. Ikhel'zon, USSR Pat. **143,848** (27 Jan 1962).
23. G. H. Beaver and G. K. Greminger, Jr., U. S. Pat. **3,226,260** (2 Feb. 1965).
24. C. Horowitz, U. S. Pat. **3,330,702** (11 Jly. 1967).
25. H. E. Hoyt and H. Pfluger, U. S. Pat. **3,498,841** (3 Mar. 1970).
26. H. Pfluger and H. E. Hoyt, U. S. Pat. **3,519,484** (7 Jly. 1979).
27. J. L. Devitl and D. H. McClelland, Ger. Offen **2,137,900** (10 Feb. 1972).
28. L. M. Gillman and R. E. Stark, U. S. Pat. **3,894,889** (15 Jly. 1975).
29. M. B. Gresham, Sb. Khim. Istochnikam Toka **9**, 161 (1974).
30. K. Sugimoto and K. Takeda, Jap. Pat. Kokai **77 38,128** (24 Mar. 1977).
31. K. Osawa, Jap. Pat. **9069** (13 Jly. 1960).
32. M. Comanor, Ger. Offen **1,094,829** (15 Dec. 1960).
33. M. Mendelsohn, U. S. Pat. **3,013,099** (12 Dec. 1961).
34. C. M. Rosser and R. A. Glinski, U. S. Pat. **3,091,554** (28 May 1963).
35. E. S. Megahed and R. W. Peters, Braz. Perdido **PI74 06,993** (13 Apr. 1976).
36. R. Vignaud, Fr. Demande **2,319,205** (18 Feb. 1977).
37. J. H. Annen, Can. Pat. **1,061,045** (28 Aug. 1979).
38. A. Himy, Silver-Zinc Battery - Phenomena and Design Principles, Vantage Press, New York (1986).
39. S. U. Falk and A. J. Salkind, Alkaline Storage Batteries, pp. 240-276, John Wiley & Sons,

New York (1969).
40. A. Fleischer and J.J. Lander, Editors, pp. 219-312, Zinc-Silver Oxide Batteries, The Electrochemical Society, Inc., John Wiley & Sons, New York (1971).
41. J. E. Cooper and A. Fleischer, Editors, Characteristics of Separators for Alkaline Silver Oxide/Zinc Secondary Batteries - Screening Methods, AF Aero Propulsion Laboratory, **300-SEP-65-773-10-216** (1965).
42. V. P. Korelova, Trudy Leningrad. Lesotekh. Akad. im S. M. Kirova, **2**, 167 (1957).
43. V. I. Sushkov, I. F. Aleksandrovich, A. Ya Rosenberg, V. P. Kim, and S. S. Gusev, Khim. Volokna, **15,** 64, (1973).
44. S. S. Gusev, E. P. Kalutskaya, I. F. Aleksandrovich, and A. Ya Rosenberg, Vysokomol. Soedin., Ser. A, **1978**, 77 (1978).
45. V. M. Irklei, K. Ya Reznik, and Yu A. Frolkov, Khim. Volokna, **1983**, 37.
46. L. T. Muus, Trans. Danish Acad. Tech. Sci., **4**, (1953).
47. R. S. Alwitt, Electrochem. Technol., **6**, 172 (1968).
48. W. P. Kilroy and C. T. Moynihan, J. Electrochem. Soc., **125**, 520 (1978).
49. A. A. Jenkins, W. C. Maskell, and F. L. Tye, J. Appl. Electrochem., **21**, 327 (1991).
50. A. Venkateswaran, J. Appl. Polym. Sci., **13**, 2459, 2469 (1969).
51. T. H. Meltzer, E. Weiss, and C. M. Garvin, J. Appl. Polym. Sci., **7**, 1691 (1963).
52. W. Bandel, Oesterr. Chemiker-Ztg., **65**, 283 (1964).
53. H. L. Pfluger and H. E. Hoyt, Rept. **NASA-CR-64607**, N65-33136 (1964).
54. C. Davila, A. ATCP, **6**, 146 (1966).
55. H. Takahara, S. Nomura, H. Kawai, Y. Yamaguchi, K. Okazaki, and A. Fukushima, J. Polym. Sci., Pt. A-2, **6**, 197 (1968).
56. M. P. Nosov, V. M. Irklei, S. G. Osinin, L. D. Nagaitseva, and T. S. Mel'nik, Khim. Volokna, **14**, 51 (1972).
57. V. M. Irklei, I. F. Aleksandrovich, S. I. Inozemtseva, N. L. Gordinets, and V. E. Nosov, Khim. Volokna, **1978**, 40.
58. M. Ya Ioelovich and Z. V. Ivulenok, Khim. Deric., **1987**, 3.
59. M. P. Nosov and B. D. Rysyuk, Acta Polym., **30**, 138 (1979).
60. B. D. Rysyuk, M. P. Nosov, and B. Kh. Yunusov, Khim. Volokna, **1985**, 15.
61. G. D. Jogleker and V. K. Batra, J. Sci. Ind. Resch., **19B**, 312 (1960).
62. T. P. Dirkse, Electrochem. Technol., **5**, 301 (1967).
63. W. P. Kilroy and L. Laughlin, Mol. Cryst. Liq. Cryst., **54**, 201 (1979).
64. M. Ya. Ioelovich, G. Veveris, and Z. V. Ivulenok, Khim. Deriv., **1984**, 56.
65. I. Kagawa and K. Sukai, Kôgyô Kagaku Zasshi, **59**, 797 (1956).
66. T. Takahashi and I. Sato, Sen-i Gakkaishi, **26**, 363 (1970).
67. S. Aravindanath, K. M. Paralikar, S. M. Betrabet, and N. K. Chaudhuri, Polymer, **23**, 823 (1982).
68. A. Sh. Goikhman and V. M. Irklei, Kompoz. Polim. Mater., **51**, 1 (1992).
69. A. Larena, M. Carmen Pena, and G. Pinto, J. Mater. Sci. Lett., **13**, 1235 (1994).
70. K. Iyer, R. Krishna, S. Sreenivasan, and N. B. Patil, Test. Res. J., **53**, 442 (1983).
71. G. Jayme and K. Balser, Papier, **18**,746 (1964).
72. G. Jayme and K. Balser, Papier, **19**, 741 (1965).
73. N. P. Matsibora, V. M. Irklei, K. Ya. Reznik, M. P. Nosov, and G. A. Stepanova, Khim. Volokna, **1990**, 30.
74. G. Jayme and K. Balser, Ind. Chim. Belge, **32**, 365 (1967).
75. G. Jayme and K. Balser, Papier, **21**, 678 (1967).

76. E. Wellisch, M. K. Gupta, L. Marker, and O. J. Sweeting, J. Appl. Polym. Sci., **9**, 2591 (1965).
77. A. Nunomura, H. Ito, A. Kasai, and K. Kamazawa, Rinsan Shikenjo Geppo, **20**, 282 (1975).
78. J. Kunze, G. Scheler, B. Schroeter, and B. Philipp, Polym. Bull. (Berlin), **10**, 56 (1983).
79. M. A. Ivanov and A. L. Kosoy, J. Appl. Polym. Sci., **19**, 2353 (1975).
80. H. Yokota, J. Polym. Sci., Polym. Lett. Ed., **21**, 285 (1983).
81. ASTM D4243-86, Am. Soc. for Testing and Mat'ls, Philadelphia, PA (1986).
82. O. A. Battista, Ind. & Eng. Chem., Anal. Ed., **16**, 351 (1944).
83. J. J. Kelley and A. J. Salkind, J. E. Cooper and A. Fleischer, *op. cit.*, p. 85.
84. E. C. Ibe and S. N. Ibe, J. Appl. Polym. Sci., **23**, 2529 (1979).
85. C. Horowitz, M. Sanduja, and P. Thettathil, U. S. Pat.**5,342,659** (30 Aug. 1994).
86. E. Kawano and Y. Nagai, Jap. Pat. **06,176,751** (23 Jun. 1994).
87. K. Ookubo, T. Kaneko, H. Iwaki, and S. Tsuchida, Jap. Pat **05,335,006** (17 Dec. 1993).
88. K. Ikeda, Jap. Pat **06,187,965-62** (08 Jul. 1994).
89. K. Imazawa and N. Takahashi, Jap. Pat. **06,275,250** (30 Sep. 1994).
90. M. Ooshima and M. Ando, Jap. Pat. **06,223,799** (12 Aug. 1994).
91. M. Tanaka, Y. Iwasaki, and H. Yamazaki, Jap. Pat. **06,196,144** (15 Jul. 1994).
92. J. You, Y. Zhan, and Z. Lin, Dianchi, **22**, 51 (1992).

ACKNOWLEDGEMENTS

This work was generously supported by the Standard Hardware Acquisition and Reliability Program at the Naval Surface Warfare Center, Crane IN, the Office of Naval Research, Washington D.C., the Naval Undersea Warfare Center, Keyport WA, and NASA-Lewis, Cleveland OH. Thanks are also due to Mr. V. Hammersley and Mr. S. Wharton of the Naval Surface Warfare Center, Crane IN for much of the analytical data, and to Mr. A. Himy, Westinghouse MTD (NAVSEA 03E21) for technical discussions.

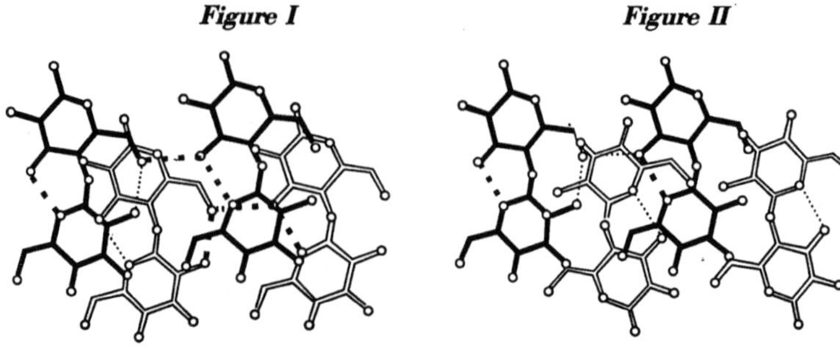

Figure 1. Parallel (I) chain structure for cellulose and antiparallel (II) chain structure for cellophane. Ref. (1).

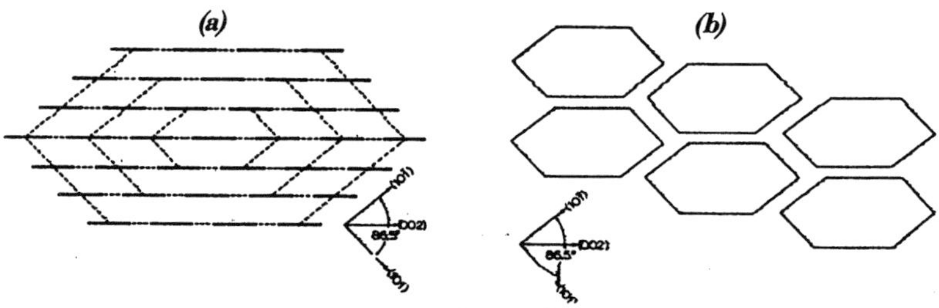

Figure 2. Microstructural units in cellulose with plane orientations. Ref. (4)
(a) Hexagonal unit microfibril configuration.
(b) Possible mode of hexagonal unit microfibril growth and layering.

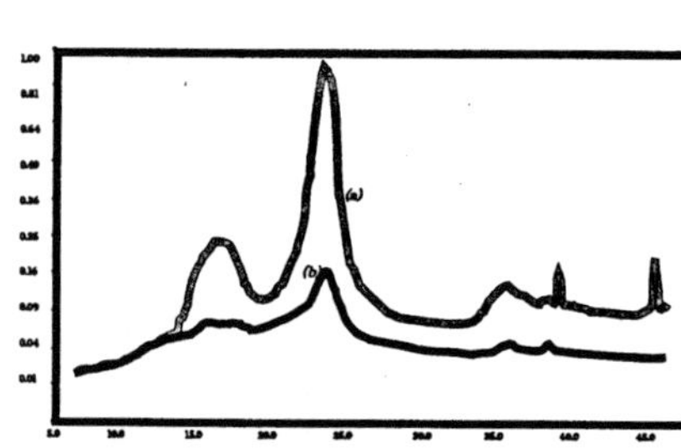

Figure 3. XRD spectra for (a) wood pulp, and (b) cotton celluloses.

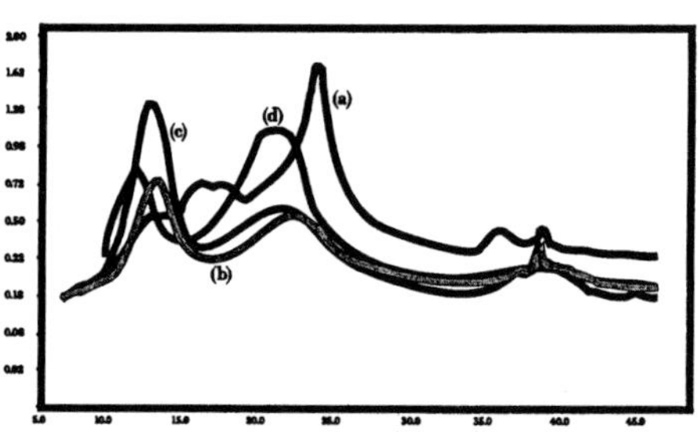

Figure 4. XRD Spectra of (a) cotton, (b) wood pulp cellophane H (c) wood pulp cellophane I, and (d) cotton cellophane F.

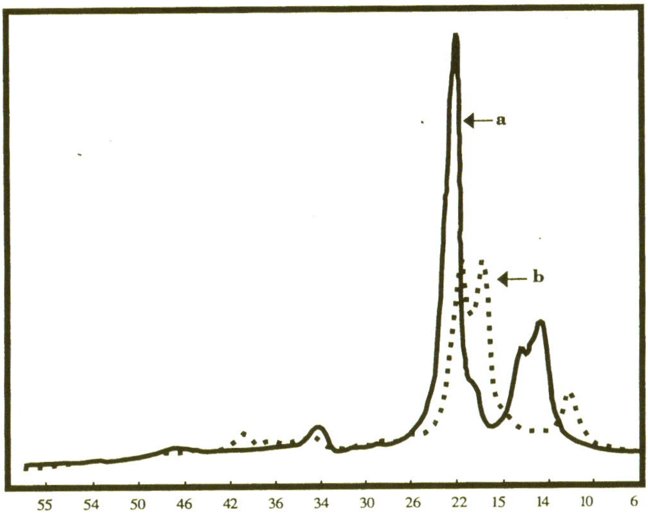

Figure 5. XRD spectra of (a) cellulose from cotton and (b) cellophane. Ref. (79).

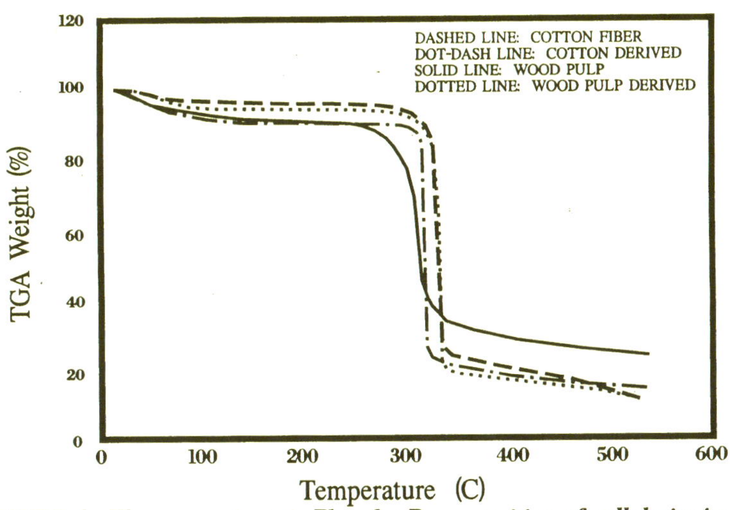

FIGURE 6. Thermogravimetric Plots for Decomposition of cellulosics including: cotton and cotton derived and wood pulp and wood pulp derived materials.

DEVELOPMENT OF SILVER-ZINC CELLS OF IMPROVED CYCLE LIFE AND ENERGY DENSITY

R. Serenyi, J. Kuklinski, D. C. Williams and A. F. Thompson
Yardney Technical Products, Inc.
Pawcatuck, CT 06379

ABSTRACT

Substantial improvements to the cycle life and energy density of silver-zinc cells have been achieved. These improvements, which translate into a longer cycle life (by 70% or better), and a significant increase in volumetric energy density, were realized through advances in the state-of-the-art of the zinc electrodes and of the separators, which are the major limitations of this electrochemical couple. To achieve these objectives, two basic approaches were employed: 1) reliance on a new class of materials, known as Electro Permeable Membranes and 2) use of bismuth oxide (Bi_2O_3) as an additive to the negative electrode in conjunction with CdO and PbO.

BACKGROUND INFORMATION

State-of-the-art silver zinc cells offer the highest power density among commercially available rechargeable batteries (up to 800 W/kg for continuous discharges and 3,000 W/kg for short duration pulses), very high energy density (up to 250 Wh/kg), a low self-discharge rate and a flat voltage during most of the discharge. However, they suffer from two serious limitations: 1) a relatively short wet life (2-3 years, maximum) and 2) a fairly rapid capacity degradation, which reduces the number of useful deep discharges to 50-100, depending on use conditions (rate of charge and discharge, temperature, etc.). These shortcomings are due to deficiencies of two of the cell's major components: the zinc electrode and the separators.

The Zinc Electrode

This component is responsible for the cycle life limitation in silver-zinc cells. The reasons are shape change and zinc dendrite formation.

Shape Change. Is a phenomenon whereby zinc oxide, formed during the discharge, is partially dissolved in the electrolyte and redeposited during recharge in a location different from where it originated. The result is a gradual depletion of the active material at the top and edges of the electrode. Attempts at reducing shape change through the use of excess zinc over the amount required for stoichiometric equilibrium with the active silver, the use of oversized negatives and the inclusion of special additives and binders have been only moderately successful.

Zinc Dendrites. Are a sharp, needlelike crystalline form of the metal, that are produced during overcharge. These dendrites have a tendency to puncture the separators, causing cell failure by internal short circuits. There are several ways to minimize the formation of dendrites: 1) careful charge control, including individual cell voltage monitoring; 2) the use of excess negative material so that the negative electrodes are not overcharged and 3) nonconventional charge methods (e.g., pulse charging).

The Separators

Regenerated cellulose, as plain or treated cellophane or fiber reinforced sausage casing, has been used as the main separator for silver-zinc cells since Professor Andre developed this system.

Indeed, it is nearly ideal in many respects, but it suffers from one serious weakness--limited resistance to oxidation by silver oxides, by oxygen, and by the electrolyte. As a result, the life of silver-zinc cells is limited to 2-3 years. Other separators that have been tried in the intervening years include grafted and crosslinked polyethylene films and microporous polypropylene. The latter offers some distinctive advantages over the cellulosics. They have lower electrolytic resistance and a high resistance to oxidative degradation and they are very thin, allowing for significant gains in energy density by creating more space for active materials. However, microporous polypropylene films alone do not prevent migration of heavy ions and molecules, such as silver, zinc or their oxides. For this reason, they must be coated for effective use in silver-zinc cells. The coating of microporous polypropylene with Electro Permeable Membranes (EPMs), in combination with improved negative electrodes developed in this program, has resulted in significantly improved wet life and cycle life.

EXPERIMENTAL

The discussion shall be limited to the most important tasks leading to technical objectives of improving the cycle life and energy density of the silver-zinc system through advances in the state-of-the-art of the negative electrodes and the separators. This has been achieved through two means: 1) reliance on a new class of materials, EPMs (1). EPMs are the analog of biological membranes which can keep two different solutions or electrolytes separated indefinitely yet can freely transfer selected ions and consist of a two-polymer blend. One of the polymers is a long chain hydrogel while the other is a supporting material and can be one of the common plastics such as polypropylene, polyvinyl chloride (PVC), or polyvinylidene fluoride (PVDF). For use in silver-zinc cells, several formulations were tested in three different capacities: a) as an additive to the negative electrode; b) as a coating to or reinforcement of the negative electrodes and c) as a coating to the separators. 2) The use of bismuth oxide, Bi_2O_3, as an additive to the negative electrodes.

The use of Bi_2O_3 as an additive to the zinc negative was previously reported by scientists at Brookhaven National Laboratory (2). The objective of this work was to evaluate its utility with other additives.

Initial studies were carried out with LR-12 cells. (12 Amp-hr capacity) These were discharged at 4.0 Amps and charged at 0.500 Amps and showed that cells using 4% EPM No. 17K added to the negative electrodes and 3 turns of EPM coated Celgard 3401 as the main separator resulted in a 69% increase in cycle life when cycled continuously at 50% Depth of Discharge with one 100% capacity discharge every fifteen cycles.

Cells containing 10% Bi_2O_3 in the negative without any other additives did not show any improvement in cycle life. Both of these studies have been previously reported (3), (4).

The second phase of this work emphasized continuing efforts to enhance cycle life through the use of EPMs and the evaluation of Bi_2O_3 in conjunction with other additives. Procedures were developed for the manufacture of composite zinc oxide and zinc electrodes with EPMs or Bi_2O_3, and for the coating of separators with EPMs. These included: 1) composite negative electrodes with solvent-based EPM which were made by mold pressing; 2) Composite zinc oxide/water based EPM electrodes manufacture was carried out by pasting which is also used for standard zinc oxide electrodes; and 3) the manufacture of zinc oxide electrodes with bismuth oxide additive which was also accomplished by pasting. All of the above methods apply to the so called "dry unformed" cells, i.e., those consisting of silver and <u>zinc oxide</u> electrodes. For the manufacture of "dry charged" cells, consisting of silver oxide and <u>zinc</u> electrodes, Yardney electroformed the zinc oxide, by electrochemically reducing the zinc-oxide from pasted strips with EPMs or Bi_2O_3. The coating of Celgard with EPMs was initially performed by hand, with rollers which seemed adequate for small scale production such as required to build the 8.5Ah cells for cell groups 1A, 1B and 2, as explained later. However, problems were experienced with groups 1A and 1B cells and, to a lesser degree, with group 2 cells which were traced to a lack of uniformity in thickness or entrapment of air bubbles in the EPM coating.

Because of those problems, Yardney has subsequently employed a subcontractor using automatic coating equipment to manufacture the separators for the group 3, 190Ah cells. Coating is done as a continuous process and its thickness is controlled to ± 0.05 mils.

Group 1 Cell Design. During the second Phase of this program, three groups of cells were designed, manufactured and tested. Each iteration incorporated the best features of the previous one. The first two groups consisted of nominal 8.5Ah cells and were built into the same 12Ah case as the Phase I cells. The control cells for both groups were identical to each other.

Group 1: The cells of this group included 33 "dry unformed" cells (3 each of 11 variations, designated as group 1A) and 27 "dry charged" cells (3 each of 9 variations, designated as group 1B). The test cells were built to retain the parameters of the controls, except those that were intentionally varied.

Tables I and II contain the most important design parameters of the group 1A (LR8.5-X) and group 1B (LR8.5DC-X) cells, respectively.

TABLE I - LR8.5-X CELLS (GROUP 1A)

Variation No.	Negative Electrode Additive (1)/Coating	Main Separator
-X1	1% PbO, 2% CdO/none	5T C-19
-X2	1% PbO, 2% CdO, 5% Bi_2O_3/none	5T C-19
-X3	Not Built	-
-X4	4% EPM 17k (2)/none	3T Celgard (4)
-X5	4% EPM 1718 (3)/none	3T Celgard (4)
-X6	1% PbO, 2% CdO/(5)	1T Celgard (4)
-X7	4% EPM No. 2518 (3)/none	3T Celgard (4)
-X8	4% EPM No. 17k (2)/none	2T Cellophane (4)
-X9	4% EPM No. 1718 (3)/none	2T Cellophane (4)
-X10	1% PbO, 2% CdO/(5)	1T Cellophane (4)
-X11	4% EPM No. 2518 (3)/(5)	1T Cellophane (4)
-X12	4% EPM No. 1718 (3)/none	5T C-19

NOTES: (1) Variations -X4, -X8 and -X12 were mold pressed; all others were pasted.
(2) Solvent based.
(3) Water based.
(4) Coated with EPM No. 1718 (one side only).
(5) Reinforced with EPM No. 1418 on a PVA paper substrate.

TABLE II - LR8.5DC-X CELLS (GROUP 1B)

Variation No.	Negative Electrode Additive (1)/Coating	Main Separator
DC-X1	1% PbO, 2% CdO/none	5T C-19
DC-X1A	1% HgO/none	5T C-19
DC-X2	1% PbO, 2% CdO, 5% Bi_2O_3/none	5T C-19
DC-X3	4% EPM No. 2518/none	3T Celgard (2)
DC-X4	4% EPM No. 1718/none	3T Celgard (2)
DC-X5	1% PbO, 2% CdO/(3)	1T Celgard (2)
DC-X6	4% EPM No. 2518/none	2T Cellophane (2)
DC-X7	4% EPM No. 2518/(3)	1T Cellophane (2)
DC-X8	4% EPM No. 2518/(3)	1T Celgard (2)

NOTES: (1) All variations were pasted and electroformed except DC-X3, which was mold pressed. All EPMs were water-based
(2) Coated with EPM No. 1718 (one side only)
(3) Reinforced with EPM No. 1418, on a PVA paper substrate

<u>Group 1 Cell Tests</u>

The 60 cells of groups 1A and 1B were tested for cycle life only. The group 1A cells "dry unformed" were tested as follows:

(1)	Fill with 23 cc's of 40% KOH		
(2)	Soak for 72 hours, minimum		
(3)	Cycle F-1	Charge:	.500A to 2.08V
		Disc.:	4.25A to 1.10V
(4)	Cycles F-2 and F-3	Charge:	.600A to 2.05V
		Disc.:	4.25A to 1.10V
(5)	Cycles 1 and 2	Charge:	.750A to 2.05V
		Disc.:	4.25A to 1.10V
(6)	Cycles 3-14, 16-29, 31-44, etc.	Charge:	.750A to 2.05V
		Disc.:	4.25A for 1.00 hr.*
(7)	Cycles 15, 30, 45, 45, etc.*	Charge:	.750A to 2.05V
		Disc.:	4.25A to 1.10V

(*) Testing is terminated when the cells are no longer able to deliver 4.25 ampere-hours.

The group 1B cells were tested the same way as group 1A, except for the procedures which are normally different for "dry charged" cells, i.e., steps (3) and (4) were omitted and the charge portion of step (5), cycle 1 was replaced with a top-off charge (.400A to 2.10V).

Group 1 Test Results. The results of the full capacity discharges for the best performing Group 1A cells and that of the standard cells (LR8.5-X1) are shown in Table III.

TABLE III - FULL CAPACITY DISCHARGES AT 4.25A - GROUP 1A

Variation No.	LR8.5-X1 (2)	LR8.5-X2	LR8.5-X12
Minimum Voltage (1)	1.501	1.501	1.505
Output (Ah) (1)			
CY 1-2	8.33	8.86	9.88
CY 15	8.85	9.11	9.21
CY 30	7.32	8.22	8.37
CY 45	6.20	6.89	6.90
CY 60	5.75	6.31	7.67 (4)
CY 75	5.22	5.85	6.33
CY 90	6.79 (3)	7.36 (3)	5.96
CY 105	4.85	6.12	5.36
CY 120	(5)	5.05	(5)
CY 135		4.72	
CY 150		4.58	
CY 165		(5)	

NOTES: (1) All data points represent the average of three cells
(2) These are the control cells
(3) Overcharged by 4.0Ah, cycle 86
(4) Overcharged by 4.0Ah, cycle 48
(5) < 4.25Ah

The average number of cycles to failure for each variation, and the corresponding failure mode for the Group 1A cells are shown in Table IV.

TABLE IV - NUMBER OF CYCLES TO FAILURE (GROUP 1A)

Variation Number	No. of Cycles to Failure	Failure Mode
-X1	111	Low Capacity
-X2	151	Low Capacity
-X4	38	Low Capacity
-X5	40	Low Capacity
-X6	16	Internal Shorts
-X7	41	Low Capacity
-X8	35	Low Capacity
-X9	31	Low Capacity
-X10	19	Internal Shorts
-X11	23	Internal Shorts
-X12	113	Low Capacity

At least one cell of each variation was dissected after failure. The findings were as follows:

Variations 1, 2 and 12: Cells of these variations presented characteristics common to aged cells, especially extensive shape change of the negative electrodes. The cells of variation -X1 and -X2 lost about 60% of the active area, mostly at the top and edges. Cells of Variation -X12 had significantly less shape change (only about 40% area loss). Otherwise, the cells were in good condition, considering their cycle life: the separators (C-19) were wet and pliable, there were no internal shorts, and the silver penetration was only moderate. Variations -X6, -X10 and -X11: all the cells of these variations experienced internal short circuits due to the fact that the heat sealed polyvinyl alcohol (PVA) bags placed around the negative electrodes opened up at the edges. None of the cells of the other variations (Nos. -X4, -X5, -X7, -X8 and -X9) had internal shorts; however, they lost capacity fast, failing to deliver 4.25Ah within 29 to 43 cycles. Although there were differences between cells of these variations, all of them shared the following characteristics: 1) extensive shape change (approximately 30% to 50% of the negative area was lost); 2) the negative electrodes were very dry, in spite of the fact that there was plenty of free electrolyte in the cells and 3) some imperfections in the coating of the separators were noted at the time of cell dissection, including partial delamination.

Group 1B Test Results. The results of the full capacity discharges for the best performing Group 1B cells and those of the standard cells (LR8.5DC-X1 with 1% PbO/2% CdO and LR8.5DC-X1A with 1% HgO) are shown in Table V.

TABLE V - FULL CAPACITY DISCHARGES AT 4.25A - GROUP 1B

Variation No. Minimum Voltage (1) Output (Ah) (1)	LR8.5DC-X1 1.496	LR8.5DC-X1A 1.494	LR8.5DC-X2 1.496
CY 1-2	11.47	11.45	11.43
CY 15	9.71	9.64	10.09
CY 30	9.05	8.94	9.75
CY 45	8.52	8.41	9.23
CY 60	7.99	7.90	8.62
CY 75	7.33	7.27	8.17
CY 90	6.46	6.40	7.62
CY 105	5.89	5.71	7.00
CY 120	5.16	4.98	6.43
CY 135	4.42	4.27	5.85
CY 150	(2)	(2)	5.41
CY 165			4.82
CY 180			4.48
CY 195			(2)

NOTES: (1) All data points represent the average of three cells
(2) < 4.25Ah

The average number of cycles to failure and the failure mode for each of the Group 1B cells are shown in Table VI.

TABLE VI - NUMBER OF CYCLES TO FAILURE (GROUP 1B)

Variation Number	No. of Cycles to Failure	Failure Mode
DC-X1	141	Low Capacity
DC-X1A	137	Low Capacity
DC-X2	192	Low Capacity
DC-X3	39	Low Capacity
DC-X4	44	Low Capacity
DC-X5	46	Low Capacity
DC-X6	58	Low Capacity
DC-X7	26	Internal Shorts
DC-X8	23	Internal Short

As with group 1A, a minimum of one cell per variation was dissected. The findings were similar to those described for group 1A. Variations DC-X1 and DC-X2. These cells performed better than the corresponding cells of group 1A by over 25%. The most likely reason for this improvement was that mixing was enhanced to ensure the uniform distribution of the additives within the negative electrodes. The reason for their eventual failure was shape change of the negatives, with a loss of active area of about 60%. However, it should be noted that the DC-X2 cells took 36% more cycles to reach that stage. The cells with 1% HgO (DC-X1A) showed no significant difference in performance, or during dissection, when compared to those with PbO/CdO. The cells with EPMs improved substantially in terms of integrity of the PVA bags and wetness of the negative electrodes, but their performance remained poor.

Group 2 Cell Design. The cells of this group were to be based on the best performing variations of groups 1A and 1B. This was done for the cells with bismuth, which were essentially the same as the corresponding group 1 cells, the only difference being the development of improved manufacturing procedures for the negative electrodes, to ensure the uniform distribution of all the additives within those electrodes.

However, in view of the poor results obtained with cells containing EPMs, these were redesigned entirely. The practice of reinforcing the negatives with EPM coated PVA paper was discontinued and positive and negative electrodes were wrapped individually with one layer of EPM coated Celgard 3401 and one layer of uncoated non-woven nylon.

As with the group 1 cells, some key parameters were kept constant, and the design of the standard cells was the same (refer to Group 1 Cell Design). Four cells of each of five variations of "dry unformed" cells, designated as LR8.5-Y, and five identical variations of "dry charged" cells, designated as LR8.5DC-Y were designed and manufactured. The most important parameters of these cells are listed in Table VII.

TABLE VII - LR8.5 AND LR8.5DC CELLS, GROUP 2

Variation No.	Negative Electrode Additive	Separator
-Y1	1% PbO + 2% CdO	5T C-19
-Y2	1% PbO + 2% CdO + 5% Bi_2O_3	5T C-19
-Y3	1% PbO + 2% CdO	1 bag of Celgard coated with EPM #1718A/1 bag of non-woven nylon (1)
-Y4	4% EPM	same as -Y3
-Y5	1% PbO + 2% CdO	same as -Y3, but with EPM #1735
DC-Y1	Dry charged version of -Y1	same as -Y1 and -Y2
DC-Y2	Dry charged version of -Y2	same as -Y1 and -Y2
DC-Y3	Dry charged version of -Y3	same as -Y3 and -Y4
DC-Y4	Dry charged version of -Y4	same as -Y3 and -Y4
DC-Y5	Dry charged version of -Y5	same as -Y5

NOTE: (1) Each plate, (+) and (-), is wrapped individually

Group 2 Cell Tests. The cells of this group were divided into two sets of 20 (2 cells of each variation per set) and tested as follows. Set A: Continuous cycling. The "dry unformed" cells, -Y1 through -Y5 were tested in exactly the same way as the LR8.5-X cells of group 1A. The "dry charged" cells, DC-Y1 through DC-Y5, were tested in exactly the same way, as the LR8.5DC-X cells of group 1B. Testing is terminated for each cell, when no longer able to deliver 60 min. at 4.25A. Set B: Characterizations tests. These consist of a series of cycles to measure the following parameters: Gas evolution on charged stand; low temperature discharge (0 and -20°C); high rate discharge (12 and 24A); high rate charge and capacity after 90 days stand.

Group 2 Test Results. The results of the continuous cycling test (set A) are shown in Tables VIII and IX.

TABLE VIII - LR8.5 AND LR8.5DC CELLS, GROUP 2

VAR.	AVERAGE OUTPUT (Ah)										
NO.	Cy1-2	Cy30	Cy60	Cy90	Cy120	Cy150	Cy180	Cy210	Cy240	Cy270	Cy300
-Y1	10.44	10.23	8.50	6.68	6.01	5.62	4.81	(2)	-		
-Y2	11.30	10.74	9.38	8.33	7.70	7.42	6.59	6.00	5.12	4.85	4.90
-Y3	10.81	9.93	8.63	7.69	7.00	-	-	-	-	-	-
-Y4	10.76	7.09	7.67	6.12	(2)	-	-	-	-	-	-
-Y5	10.34	9.90	9.59	8.67	8.23	(2)	-	-	-	-	-
DC-Y1	10.77	10.39	7.43	6.42	5.83	5.30	(2)	-	-	-	-
DC-Y2	11.59	10.92	9.39	8.27	7.42	6.99	6.14	5.72	(1)	-	-
DC-Y3	10.47	9.99	9.72	8.55	8.35	7.27	(1)	-	-	-	-
DC-Y4	10.00	5.97	4.68	(2)	-	-	-	-	-	-	-
DC-Y5	10.22	10.15	(1)	-	-	-	-	-	-	-	-

NOTES: (1) Slow short
(2) Low capacity (< 4.25 Ah)

TABLE IX - NUMBER OF CYCLES TO FAILURE (GROUP 2)

Variation Number	Avg. Number of Cycles to Failure	Failure Mode
-Y1	199	Low Capacity
-Y2	> 325 (still cycling)	-
-Y3	122	Low Capacity
-Y4	102	Low Capacity
-Y5	134	Low Capacity
DC-Y1	167	Low Capacity
DC-Y2	238	Slow Short
DC-Y3	160	Slow Short
DC-Y4	63	Low Capacity
DC-Y5	43	Slow Short

The characterization tests (set B cells) are still under way, but some of these have been completed. Gas evolution tests were performed during cycle 3, with the cells in the charged condition, and stabilized at 37.8°C (100°F). The gas evolution was measured hourly during the last 8 hrs. of the 24 hr. stabilization period. The results are shown in Table X.

TABLE X - GAS EVOLUTION, CHARGED, AT 100°F

Variation No.	Avg. gassing rate (cm^3/hr)	Avg. gassing rate (cm^3/hr) ÷ gZn)
-Y1	2.50	0.096
-Y2	1.26	0.048
-Y3	0.83	0.032
-Y4	4.15	0.160
-Y5	0.94	0.036
DC-Y1	3.05	0.117
DC-Y2	1.35	0.052
DC-Y3	1.02	0.039
DC-Y4	4.93	0.190
DC-Y5	1.07	0.041

Low temperature discharge tests were done at 4.25A during cycles 4 (0°C) and 12 (-20°C). The results indicate no significant difference in capacity of voltage between variations, except for -Y4 and DC-Y4, which were significantly lower. High rate discharge tests were done at 12A (cycle 6) and 24A (cycle 8). Again, there were only minor differences between variations, except for -Y4 and DC-Y4.

EVALUATION OF THE RESULTS

Bismuth Oxide

The beneficial effects of this additive did not materialize until the second phase of the program when it was found that a mixture of Bi_2O_3 with PbO and CdO would enhance the cycle life of the cells, by a synergistic effect. Improvements to the manufacturing process, introduced during the construction of the group 2 cells, further added to their cycle life, to the point that these cells exceeded 325 cycles.

A comparison of the life of these cells and the controls, at various stages of the program, shown in Table XI is self explanatory.

TABLE XI - NO. OF CYCLES TO FAILURE

Test Group	Controls	Cells w. Bi_2O_3	Δ
First Phase	96 (1)	< 10 (2)	N/A
Second Phase			
Group 1A	111	151	+36.0
Group 1B	141	192	+36.2
Group 2, dry unformed	199	325[+]	+63.3[+]
Group 2, dry charged	167	238	+42.5

NOTES: (1) With 1% HgO. All other control cells with 1% PbO and 2% CdO.
(2) With 10% Bi_2O_3 and no other additive.
All others with 5% Bi_2O_3 and PbO/CdO.

In addition, the cells with Bi_2O_3 generate hydrogen gas at only about 50% the rate of the controls, as evidenced by their superior cycle life. Their high rate and low temperature performance are equivalent to that of the controls.

EPMs

The results with EPMs were excellent at times, but lacked consistency because of several problems experienced during this program. In summary, the following was observed: 1) During Phase I, one variation of LR12 cells, with EPMs in the negative electrode (4%, solvent based), and 3 turns of coated Celgard, outlasted the standard cells by over 68% (160 cycles vs 95), and showed better capacity through testing. Thus, by cycle 90, their capacity was 70% higher. 2) Group 1A cells of Phase II did poorly, with just one variation (-X12) outperforming the standard cells, and only by the narrowest of margins. The group 1B cells showed significant improvements over the corresponding cells of group 1A, but their performance remained below expectations. 3) Dissection of group 1A and 1B cells identified the reasons for this disappointing performance: a) the heat sealed, EPM coated polyvinyl alcohol (PVA) bags opened up at the edges, exposing the cells to premature internal shorts; b) excessive dryness of the negative compartment; and c) imperfections in the coating of the separators, including voids and partial delamination.

As a result of these problems, the group 2 cells with EPMs were completely redesigned, as described under "Group 2 Cell Design". In summary, EPMs were no longer used as additives to the negatives, but only as a coating to single layers of Celgard microporous polypropylene, used to wrap all the positive and negative electrodes. The results of group 2 cells place the EPMs back in contention in the effort to improve the cycle life and energy density of the silver-zinc cells. As seen in Table VIII, certain variations outperformed the controls by a significant margin, e.g., the cycle 90 results show:

1. Dry unformed cells

-Y1 (control)	:	6.68Ah
-Y3 (w. EPMs)	:	7.69Ah (+15.1%)
-Y5 (w. EPMs)	:	8.67Ah (+29.8%)

2. Dry Charged cells

DC-Y1 (control)	:	6.42Ah
DC-Y3 (w. EPMs)	:	8.55Ah (+33.2%)

The only remaining problems are imperfections to the coating of the Celgard, which causes the cells to short, or allows the zinc or zinc oxide to migrate outside of the Celgard bag, resulting in rapid capacity losses. We believe this problem has been solved through the use of a continuous coating process, capable of uniform (± 5% in thickness) and bubble free coatings. This material is being used in group 3 cells, presently under manufacture.

CONCLUSIONS

Bismuth oxide, used as an additive to the negative electrode in conjunction with lead and cadmium oxides, has been demonstrated to extend the life of silver-zinc cells by at least 35%. Recent work, which includes not only cells tested on this program, but also 200Ah cells for another program, points to even more significant improvements, in the order of 40 to 65%, through improved manufacturing procedures which ensure the homogeneous dispersion of the additives through the body of the negative electrodes.

The outlook for EPMs, has improved considerably and there is every reason to believe that the 190Ah cells being assembled at this time will also show improved cycle life.

It must be noted that EPMs, when used as a coating to the separators, as in the case of the above 190Ah cells, allow for less separation between adjacent plates, thus creating more room for active materials and therefore, increased energy density by at least 15%.

ACKNOWLEDGMENT

The authors wish to acknowledge the support and assistance of Dr. Patricia Smith, Dr. Stanley James and Dr. Charles Fleischmann, Naval Surface Warfare Center, Silver Spring, Maryland. This work was carried out under Contract N60921-91-C-00205 and Contract N60921-93-C-0062, funded by SBIR Programs under the Sponsorship of the Office of Naval Research

REFERENCES

1. R. L. Peck, Ionic Semiconductors Materials and Applications Thereof, U.S. Patent Number 4,797,190, Jan. 10, 1989.

2. J. McBreen and E. Gannon, "Bismuth Oxide as an Additive in Pasted Zinc Electrodes", Journal of Power Sources, 15, 169 (1985).

3. R. Serenyi, Report No. NSWCDD/TR-92/296, "Development of Silver-Zinc Cells of Improved Cycle Life and Energy Density", edited by Dr. S. James, (1994).

4. R. Serenyi, "Development of Silver-Zinc Cells of Improved Cycle Life and Energy Density", Proceedings of the 36th Power Sources Conference, 278 (1994).

DESIGN PARAMETERS AND ADVANCES IN SILVER ZINC SECONDARY BATTERIES

Charles Grun
Rutgers University, College of Engineering, PO Box 909 Piscataway, NJ 08855
Harlan Lewis
Materials Science Branch, NSWC, Crane, IN 47522
Alvin J. Salkind
UMDNJ-Robert Wood Johnson Medical School, 675 Hoes Lane, Piscataway, NJ 08854
and Rutgers University, College of Engineeering.

ABSTRACT

Silver-Zinc Secondary Batteries have become a mature industry from initial development in the early 1940's. Progress, applications, design and future advancements are discussed.

Since the introductory work of Henri Andre in the early 1940's of a practical rechargeable Silver-Zinc cell, there have been many applications of this energetic and safe system. Silver Zinc batteries have been used in satellites such as Sputnik 1, Explorer XII, submarines Albacore, Dolphin, NR-1, torpedoes, torpedo targets, deep submergence rescue vessels, lunar spacecrafts, mundane items like intense beam flashlights, high value portable electronic and medical equipment.

Those of you who have seen the movie Apollo 13 may be interested that the battery in the Lunar Excursion Module (LEM) was a Silver Zinc Design. The LEM was where the three astronauts hunkered down after their oxygen tank exploded, 3 days into their space flight, putting their service module and bulk of fuel cells out of commission. They used the LEM, powered by Silver Zincs, essentially as a lifeboat to swing around the moon and for most of the trip back to earth. That Silver Zinc battery, of which there were three aboard, had a specification requirement of 5 cycles, capacity of 296 ampere-hours (AH), a nominal voltage of 30 volts, and a maximum weight of 120 lbs including the appreciable weight of the battery container, Figure 1. This comes to a specific energy of 162 watt-hrs/Kg or 73.6 watt hours/lb. It was also very compact with a high volumetric energy density of 356 watt hours/Liter (5.83 watt hours/ in^3).

Let's look first at some of the generic applications of the Silver Zinc battery,design parameters and then at problem areas that still bother the battery and what is being done about them.

In general, the Silver Zinc cell is mainly limited in its use not by technology, but by cost, to a reliable high energy and power density supply. At the price of silver @ $5.3/troy ounce, its cost per AH is 51 cents, based on a practical 3 gm Ag/AH. (Theoretical utilization is 2 gm Ag/AH). The manufacturing cost of the Silver Zinc cell is also high, since the market is limited and within this market there is a profusion of cell sizes which severely restricts automatic manufacturing processes.

Depending on size, discharge rate and life requirements, large size Silver Zinc cells (greater than 100 AH) can be designed with specific energy in range of 230 watt hours/Kg and volumetric energy density of 615 watt hours/Liter, (106 watt-hours/lb and 10 watt-hours/in^3).

Design Parameters.

What does the Design Engineer have in his portfolio to design Silver Zinc cells? As mentioned, Silver Zinc cells, because of their cost, are application specific. Which means that the cells that make up the battery have to meet requirements, such as capacity at specified discharge rates within voltage limits, specified ambient temperatures, cycle or calendar life, and a host of environmental conditions such as vibration, shock etc.

The life cycle limitations of Silver Zinc cells can be attributed to the solubility of both electrodes,(silver oxide and zinc) and changes in morphology in electrode structures on charging. As documented by Andre, electrolyte excess can greatly accelerate the deterioration of the electrode structure.

Separator

The basic and probably the most contentious choice is the separator system , since this is most vulnerable to the ravages of the electrolyte and the soluble products of the electrodes. The main practical separator of the system is cellophane, with interseparators such as woven nylons, and felt-like propylenes, acrylonitriles and nylons used directly on the electrodes as spacers or electrolyte reservoirs. The purpose of the cellophane membrane is to retard silver migration to the zinc electrode, which in its absence would short the cell. It also retards the zinc penetration from the zinc side to the silver, caused by growth of zinc dendrites during cycling.

Cellophane is a semipermeable membrane, derived from cellulose (trees, cotton etc) and manufactured, so far, only by the viscose process, which involves cellulose dissolution in NaOH, xanthation, extensive filtration and eventual coagulation in a sulfuric acid - zinc sulfate bath, followed by washing and drying. Its pore size is reported to be in the 20 Angstrom range

For maximum strength and least degradation, a high degree of polymerization (DP) and crystalinity is desired. However, the viscose process, driven mainly by the requirements of the packaging industry, severely degrades these and even starts (in the US and England) with inferior cellulose, trees, instead of high DP cotton.

Table l lists some of the characteristics of cellophane battery separators.The choice is not large. The best in the strength column are the Russian cellophanes, presently not available, since the factory in Kazan that produces it is now closed. The strength probably goes hand in hand with the least swelling. The starting material of the Russian cellophane was cotton. For the other cellophanes the starting materials were trees.

The cellophane, as manufactured, can be given an improved silver retardation capability and greater oxidation resistance by treatment with a water soluble silver salt [(1)]. This silver salt apparently oxidizes the cellulosic terminal aldehydes to carboxyl groups, which have a greater resistance to further oxidation [(2)]or degradation

Besides the cellophanes, there are grafted polymeric membranes available for use in Silver Zinc cells. The base film is generally a 1 mil polyethylene, modifed by adding side chains to impart lower electrical resistance [(3)] In general, these separators have a higher electrical resistance and are much more expensive than cellophane. Cellophane costs about 2.5 cents/ft^2, while grafted membranes range in the area of \$3.00/ft^2. If employed, they are usually used in conjunction with cellophane, i.e. several layer of cellophane plus 1 layer of the grafted polymeric membrane.

To avoid membrane contact with the highly oxidizing silver electrode, an interseparator or basically a spacer consisting of nylon woven cloth or non-woven felt is used. On the zinc electrode a non-woven Dynel (acrylonitrile) or nylon felt is generally used, both as a "spacer" and electrolyte reservoir.

The cellophane separator is somewhat unique in that it swells in thickness about 200% when wet(from 1 mil dry to 3 mil wet). This allows for easy insertion of dry cell pack into cell case, and constrains the electrodes when wet with electrolyte. Wet design pressure is in the neighborhood of 4 psi.

The electrodes may be wrapped with the separator in several ways. For a "normal wrap" (NW), two silver electrodes with bottoms facing each other are wrapped with the separator. The separator is then folded at the center-line (bottom of electrode) to form a U type envelope. The zinc electrodes are then placed at the center and on either side of the U. For a "reverse wrap" (RW) the zinc electrode is placed inside the envelope with the silvers on the outside of the wrap. For a "split wrap" (SW), the silver electrode is inside the bag but with only a partial amount of separator. The remaining separator is wrapped on single zinc electrodes with the bottom part of the separator folded or shut with a clip. No absolute advantage is evident for either wrap. The NW is most economical, since rugged silver electrodes are wrapped (handled). The RW may minimize zinc shedding but is more prone to bottom and side splitting on extensive cycling if the zinc does shed, or cell is overcharged. The SW is more labor intensive and may have extra pressure at bottom, due to separator overlap, if not properly designed or assembled.

Electrolyte.

Potassium hydroxide (KOH) is the electrolyte used in the Silver Zinc cell. The concentration is in the range of 38 to 45%. Lower concentrations are used for higher voltage and low temperature operation. Higher concentrations are however less deteriorating to the cellophane separator.

Additives are at times added, but rarely in the US. The electrolyte used in Russian Silver Zinc cells contains zinc oxide, lithium hydroxide, and sodium tetraborate. The zinc oxide reduces the zinc electrode dissolution, the lithium hydroxide "stabilizes" the zinc electrode preventing passivation and the sodium tetraborate stabilizes the cellophane, i.e. decreases swelling to maintain strength (4).

Many electrolyte additives have been studied. A compilation of this work is presented by Bass (5) and more recent work is presented by Shivkumar (6). The additions range from lead oxide, through phosphates, flourides, borates, quarternary ammonium salts, thiurea, surfactants and gelling agents. It is of interest that none are being used in secondary Silver Zinc cells by US commercial vendors. The problem may be that the "additive" studies involved only several parameters, such as zinc capacity fall-off at specific cycling, while a commercial vendor has to check out all posible conditions, from short time activation to long term cycling, charged wet stand, wet life and a host of operating conditions.

Silver Electrode.

Silver electrodes are generally produced by direct formation of silver powder in the desired sheet structure and then sintering @ 400 to 600°C.This is accomplished by either filling a mold containing the electrode grid with a weighed amount of silver, compressing this and sintering, or continuously depositing silver powder on a paper web, adding a mesh (expanded silver), compressing this between pressure rollers, stripping away the paper web, and sintering the silver powder sheet. Variations of this method are a) resin bonded techniques, where silver powder is mixed with a plastic resin to form a composition that can be rolled into a sheet, which is then heated to remove the resin and sintered b) Ag_2O pasting method, where monovalent silver oxide with some water and small amount of carboxy methyl cellulose (CMC) is applied as a paste on a mesh, generally expanded silver,then dried and sintered. The sintering process decomposes the Ag_2O to silver.

High utilization silver powders are in the range of 1.5 to 3.0 microns. But utilization is not the only concern. The powders must not shrink excessively on sintering and must be easily spreadable, if electrodes are manufactured in the dry version. Silver electrode densities range from 3.5 to 4.8 gm/cm^3, corresponding to porosities of 50 to 67%. The higher density is more common, since more capacity can be placed within a volume.

For dry charged cells, the silver electrodes are charged in KOH against inert counter electrodes, thoroughly washed with water and then dried.

Zinc Electrodes.

Zinc electrodes are constructed by pressing dry ZnO powder onto a mesh within a mold, by continous wet slurry method, or by electroplating zinc on electrode mesh.

The zinc oxide powder contains a small amount of binder such as CMC or fibrilated teflon and additives such as lead and cadmium to reduce hydrogen evolution. To provide improved structural integrity, the electrode is wrapped and pressed into Viskon or Aldex paper which becomes an integral part of the electrode. For the slurry process, the wet zinc oxide is applied continuously on a paper web, and after drying is pressed on to a mesh. The paper remains on the electrode, serving again as reinforcement and interseparator.

Historically, mercury was added to the zinc electrode to suppress hydrogen evolution from the zinc suface. Since mercury has become unacceptable for environmental reasons, PbO and Cd (1% PbO + 2%CdO) has been substituted based on the evaluation of Himy and Wagner[7, 8]. This substitution is also supposed to have a positive effect on the zinc electrode by decreasing its shape change on cycling.

Again, many additives to the zinc electrode have been studied [9]. These range from lead oxide, cadmium oxide, thallic oxide, bismuth oxide, calcium hydroxide, acetylene black [10], titanate fibers and polytetraflouroethylene (Teflon) fibrils. The latter two are mainly mechanical stability enhancers. With the exceptions of PbO and CdO as mercury substitutes and some use of fibers for mechanical stability, the commercial community has not applied any of the above in secondary silver zinc cells.

The theoretical ratio of zinc to silver is 1.2 : 2 or 0.6. Depending on number of cycles required, this ratio may go up to 1.2, i.e. greater number of cycles will require a higher ratio. Also, while the theoretical silver utilization is 2 gm/AH, the practical design utilization ranges at 3-4 gm/AH for secondary cells.

Cell Containers

Cell cases and covers for the silver zinc cells are generally molded from styrene-acrylonitrile (SAN), acrylonitrile-butadiene-styrene (ABS) copolymers, nylon, polyphenylene oxide, or polysulfone. The choice of plastic is mainly governed by anticipated temperature. The above copolymers have the lowest and polysulfone the highest temperature resistance.

Discharge Shape, Gravimetric and Volumetric Utilization

Figure 2 shows a series of actual discharges of a Silver Zinc cell. The discharges are for a Russian SZ-70 cell with an initial low rate capacity of 120AH. At the start of discharge we see the characteristic AgO voltage which is higher and longer lasting at low rates than at higher rates. This cell has an electrode surface area of 2,410 cm^2 (374 in^2); therefore at 100 amperes we are discharging at 41ma/cm^2 (267 ma/in^2). At a wet cell weight of 1310 gm, capacity of 120AH (@40Amps) and 1.48 volts, we are looking at an initial cell rating of 136 watt-hours/Kg (61.8 watt-hours/lbs)

Figure 3 shows the gravimetric utilization of a low rate 200AH Silver Zinc Cell. The active materials account for 52% of the cell weight, electrolyte 25% and remainder 23%.

The volumetric utilization of this 200AH cell is shown in Figure 4. Electrolyte here occupies 43% of the space, active materials 16%, empty space (air) 16% and remainder 26%. The obvious reason for the low active materials volume is their high density (AgO @ 7.44 and Zn @ 7.14 gm/cc).

Advances:

While Henri Andre experimented in the 1940's with small secondary Silver Zinc Cells, this activity has now matured to a multimillion dollar industry. What are some present problems and what advances are in the horizon.?

Separator - There is always a quest to improve the Silver Zinc cycle life. Cellophane, the main player in this search, instead of showing signs of improvement, seems to have deteriorated in the last several years. The causes appears to be the increased use of inferior starting material, trees, and the fact that battery grade cellophane accounts for only a small percentage of the market, the bulk being consumed by the packaging industry.

An educational process is now under way to apprise the cellophane manufacturers on the technical requirements of battery grade cellophane for the Silver Zinc cells.

As part of environmental restrictions, there is a movement afoot to change the production method of cellophane with simplified solution procedures. This should decrease the deterioration of the DP and provide a better grade cellophane.

Also for value intensive applications, the use of grafted membranes in conjunction with cellophane should be investigated to a greater extent.

Additives - Zinc electrode and electrolyte additives should be evaluated in depth. Of the many seemingly promising additives almost none have been commercially adopted, probably because studies were not in depth or not widely application specific.

Charging Techniques - Pulse and asymmetric charging techniques may be beneficial for the Silver Zinc cells. Asymmetric charging studies conducted in the late 1960's on the LEM Silver Zinc battery were successful in eliminating the AgO voltage hump while maintaining better capacity than regular DC charging. However, the cycling was short term, ~ 5 cycles. With the much simpler electronic pulse charging technology available now, a thorough matrix of this method should be evaluated..

References

1. Mendelsohn M. US Patent 3,013,099 December 12, 1961

2. Post R. E. "The Nature of Cellophane as Related to its Application as a Zinc-Silver Oxide Separator" Zinc-Silver Oxide Batteries, Fleischer A. et al Ed. 1971, p263-269, John Wiley & Sons

3. D'Agostino et al. "Grafted Membranes" Zinc-Silver Oxide Batteries, Fleischer A. et al Ad 1971, p271-294, John Wiley & Sons

4. Bagotsky V. S. Personal Communication

5. Bass K. et al "Methods for the Reduction of Shape Change and Dendritic Growth in Zinc-Based Secondary Cells" Journal of Power Sources. 35,3 August 1991, p333-352

6. Shivkumar R. et al. "Effect of Additives on Zinc Electrodes in Alkaline Battery Systems" Journal of Power Sources, 55 1995, p53-62

7. Himy A. et al "Substitutes for Mercury in Alkaline Zinc Batteries(ll)" Proceedings 28th Power sources Symposium, Atlantic City, June 1978 p167-169

8. Wagner O. et al "Substitutes for Mercury in Alkaline Zinc Batteries" Proceedings 27th Power Sources Symposium, Atlantic City, June 1976 p135-138

9. Bass K. et al "Methods for the Reduction of Shape Change and Dendritic Growth in Zinc-Based Secondary Cells" Journal of Power Sources 35, 3, August 1991 p333-352

10. Frackowiak E. "Modification of Secondary Zinc Electrode" Power Sources #13 17th International Power Sources Symposium, April 1991, Bournemouth UK p225-235

Table 1 Characteristics of Cellophane Type Separators

Separator	Thickness Dry mils	Thickness 45% KOH 4psi, mils	%Change Width (1) 45% KOH	% Change Length (1) 45% KOH	Resistance Ohm-cm	Tensile Strength MD(2) 45% KOH	Tensile Strength TD(2) 45% KOH
Wood Pulp A	1	2.78	7.2	-1.9	10.4	1.2	0.6
Wood Pulp H Clr	1.1	2.78	-0.2	-2.7	8.9	0.91	0.44
Cotton F Clr	1.1	2.13	7.2	0.8	12.2	2.1	1.7
Cotton E, Ag	1.1	2.26	10.1	-1.4	12	2.1	1.8
Wood Pulp I	1	2.78	4.7	-3.4	8.9	1.8	
Wood Pulp J Rein.	3.3	6.65	4.7	-4.7	11.2	(3) 5.	(3) 3.
Wood Pulp L	3	6.8	6.3	-6.2		(3) 4.	3.3

(1) Length is Machine Direction. Width is Transverse Direction.

(2) MD = Machine Direction. TD = Transverse Direction. Tensile Strength is for 0.25" strip, in lbs

(3)Wood Pulp J Rein. thickness is 3.3 mil, dry. Wood Pulp L is 3 mil, dry. All other separators are ~ 1 mil, dry.

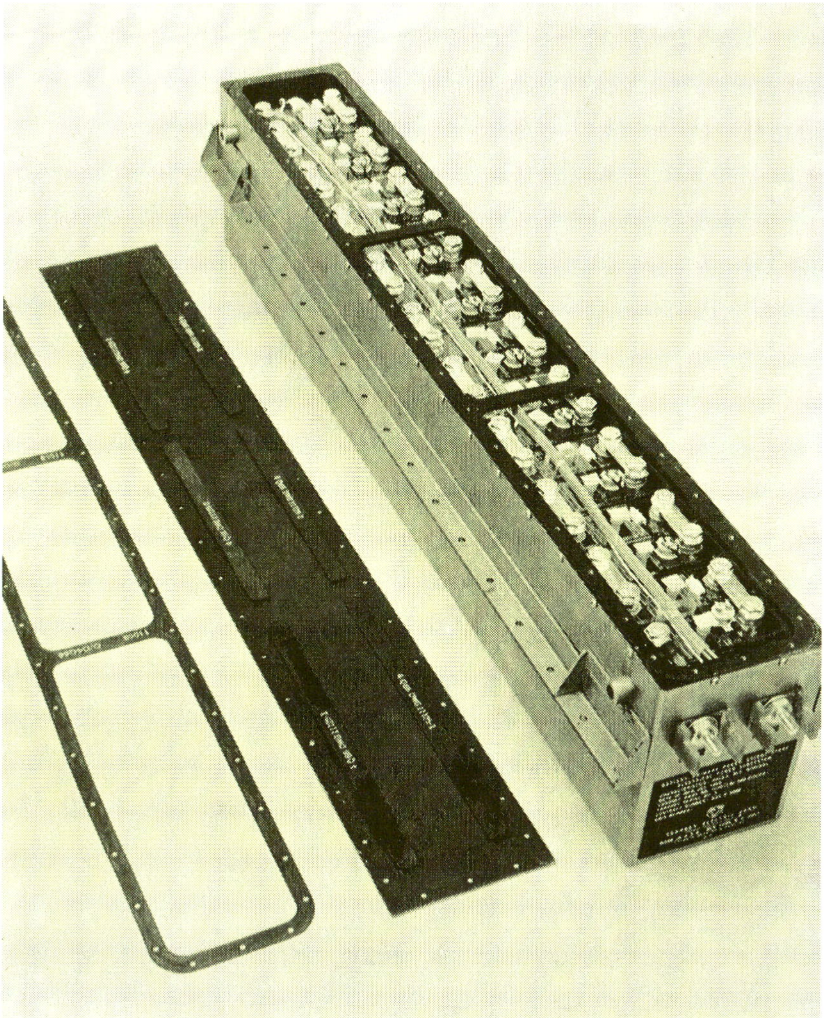

Figure 1
Lunar Excursion Module (LEM) Silver-Zinc Battery

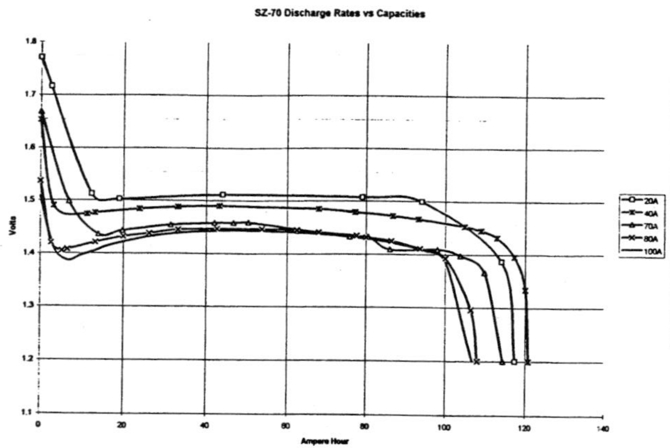

Figure 2

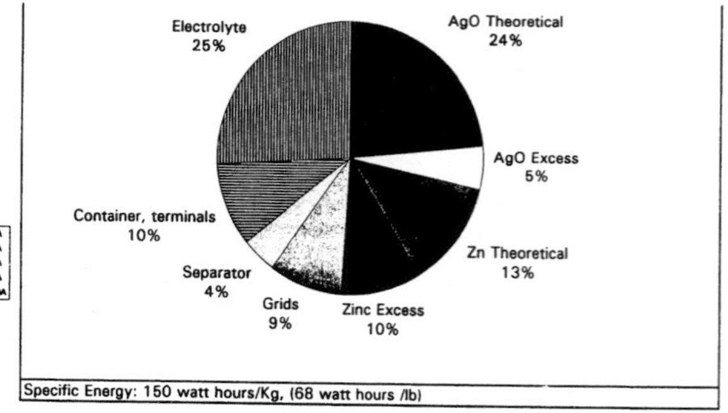

Figure 3

Gravimetric Utilization of Silver Zinc Cell (Low rate, 200AH)

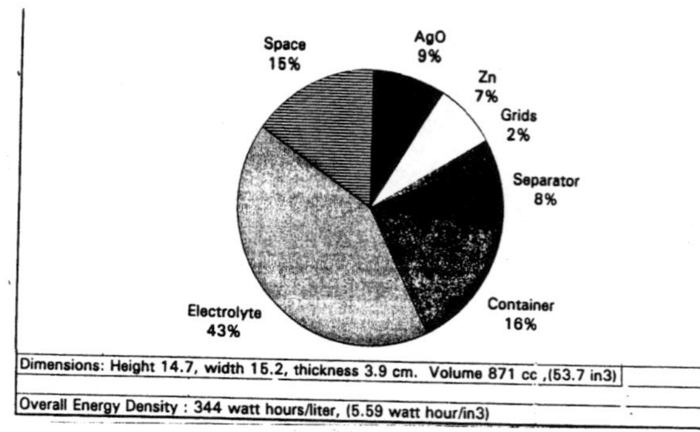

Figure 4

Volumetric Utilization of Silver Zinc Cell (Low Rate 200 AH)

MERCURY- AND LEAD-FREE RECHARGEABLE Zn/MnO_2 CELLS

Josef Daniel-Ivad[*] and Jean-Yves Huot[**]

[*] Battery Technologies Inc., 30 Pollard Street, Richmond Hill, Ontario L4B 1C3
[**] Noranda Advanced Materials, c/o Centre de Technologie Noranda, 240 Hymus Blvd, Pointe Claire, Quebec, H9R 1G5

Zinc powders were evaluated in Rechargeable Alkaline Manganese (RAM) cells with the target of eliminating lead from mercury-free zinc alloy powders while maintaining electrical performance and low gassing rates. Discharge performance at a 92% level compared to control cells with Zn-Pb powders and acceptable gassing rates were demonstrated.

INTRODUCTION

Rechargeable zinc electrodes for Ni-Zn, Zn-Air, Ag0-Zn and Zn-Flow systems are well known and the characteristics are reported in the literature (1). The zinc electrode in these systems is typically a plate electrode in the discharged state made from a zinc oxide slurry with some plastic bonding agents on a carrier grid. Zinc electrodes for Rechargeable Alkaline Manganese (RAM) round cells use a gelled zinc powder electrode that is assembled in the charged state similar to primary alkaline cells, with essentially the same shelf life and charge retention as the shelf life of primary alkaline cells (2,3).

Mercury-free Rechargeable Alkaline Manganese (RAM) cells were introduced in the consumer market in 1994. Alloying elements such as lead and indium are added to the mercury-free zinc powder, because elimination of mercury in the zinc electrode impacts on gassing and electrical performance (4). Old prior art revealed that the addition of indium to the zinc powder or to the gel may reduce zinc gassing (5). Since then many patents on zinc and zinc alloys issued worldwide.

This work was devoted to the elimination of lead in mercury-free zinc powders. Our target was a RAM cell with electrical performance and gassing close to that of cells with lead containing zinc powders. Table 1 shows the electrical performance on selected application tests for RAM AA cells available on the North American consumer market under the trade name *RENEWAL* (Rayovac, Wisconsin, USA) and *PURE ENERGY* (Pure Energy Battery Corporation, Nova Scotia, Canada).

EXPERIMENTAL

Battery grade zinc powders were made by gas atomization of 99.997 wt.% molten zinc containing various amount of alloying elements. Particle size distribution was controlled by adjusting gas atomization and classification. Physicochemical properties of those samples were evaluated, namely particle size distribution, density, chemical composition, gassing and anodic discharge. Particle size distribution was calculated from log-normal distribution of sieving analysis, apparent density and flowability were measured with Hall flowmeter, zinc gassing was measured by gas collection, and out-of-cell anodic discharge was measured in a Plexiglas electrochemical cell containing excess free KOH (4). Powders tested included pure zinc, Zn-Pb, Zn-In, and Zn-Pb-In alloys shown in Table 2.

These zinc powders were evaluated in mercury-free RAM AA cells. Figure 1 shows a cross sectional view of a RAM AA cell. The positive electrode consists of three cathode rings (pellets), which are formed under high pressure to a slightly oversized diameter and then inserted into the nickel plated steel can. A standard cathode formulation was used for the experiments consisting of electrolytic manganese dioxide, graphite and additives that improve cycle life and that facilitate hydrogen recombination (6,7). Gelled zinc anodes were prepared from the zinc powders as received using a BTI proprietary method that involves deposition of indium on the zinc particle surfaces (8). The separator used was a laminate of a non-woven material and cellophane. A brass nail functioned as negative current collector. Cathode, separator and electrolyte formulations were held constant and the zinc powder was the only variable in the experiments. The active material was chosen so that the zinc electrode was limiting the capacity of the cell.

RAM AA cells were tested for cycling performance, capacity retention, shock resistance, gassing and leakage at various states of charge. During high temperature storage cells were inspected on a weekly basis for bulging and leakage. After high temperature storage cells were removed, punctured and the amount of gas released from the cell was measured. High temperature storage of 2 weeks at 65°C was used for baseline comparisons. It was shown in previous work that 1 week at 65°C is approximately equivalent to a storage period of 1 year at ambient temperature between 15 to 35°C (3). Electrical cycling was performed on computerized test equipment utilizing a built-in 12 hour taper charge method (charge to 1.65 V). The shock sensitivity is an important parameter in mercury-free gelled zinc powder electrodes. Due to the absence of mercury, which functioned as a conductive matrix between zinc particles, the gelled zinc powder electrode became much more sensitive to mechanical impact. A drop test was carried out which involved a triple drop of RAM AA cells from a height of 1 meter onto a concrete base. First cells were dropped on the positive end (oriented in a tube), then on the negative end (oriented in a tube), and finally on the side (rolled off a ramp). Before and immediately after the drop open circuit voltage (OCV, no load) and closed circuit voltage (CCV) at a load of 1 Ampere @ 200 milliseconds were measured. Within 5 minutes after the triple drop the Motor & Toy test was started to measure capacity of the dropped cells.

RESULTS AND DISCUSSION

Properties of zinc powder

Figure 2 shows that particle size distribution (PSD) of zinc powder is well described by a log-normal distribution, which is expressed in terms of mean diameter and width. This distribution depends upon atomization conditions and classification by screens. In this work the term "fine fraction" or "fines" relates to zinc particles in the range of ASTM sieve size 200 and 325 mesh or 75 to 44 microns. A low fine fraction was defined with up to 3 wt.% of the zinc powder falling in this range and more fine fraction requires that at least 4 wt.% fall in this particle size range. The presence of coarses (larger than 600 microns) and fines increases the width of PSD. Clipped powder containing a small amount of fines display a typical width distribution of 1.5, while powders with more fines display a width ranged from 1.5 to 1.8 (Table 2). The shape distribution shows that zinc particles are elongated with an average length:width ratio of two. A small fraction of zinc particles are very elongated with a ratio higher than five.

Density of zinc powder is largely determined by the particle size distribution, but it is less influenced by its chemical composition. Zinc powder gassing is markedly inhibited by the addition of lead, while the addition of 500 ppm indium to pure zinc does not reduce the high gassing rate of pure zinc. The presence of fines has a minor impact on gassing.

Out-of-cell performance under continuous discharge (100 mA/g Zn) of gelled zinc anode is summarized in Table 2. Figure 3 displays typical discharge curves of Pb-Zn and pure zinc. The capacity was measured at -1.1V (vs. Hg/HgO) and the slope of the plateau was calculated in the linear section of the curve. Performance is strongly influenced by the chemical composition of zinc powder. The capacity increased as follows: Zn < Zn-In < Zn-In-Pb < Zn-Pb. The presence of fines did not improve capacity in out-of-cell tests.

Performance and gassing of RAM AA cells

Table 3 shows a comparison of electrical performance and gassing data of RAM AA cells with different zinc powder samples. Overall performance is best with zinc alloys containing 500 ppm lead, in agreement with the out-of-cell continuous discharge. The presence of fines improved performance of all zinc powders in RAM AA, whereas no such effect was noted in the electrochemical cell. However, gelled powders with more fines showed higher anode gassing over several weeks. This is also opposed to the gassing behaviour of powder over 24 hours. This disagreement between the effect of fines on powder and gelled powder gassing is likely due to the large difference in the time scale of gassing monitoring. In other words, gassing of fines would appear after a long exposure to concentrated KOH solution, and accelerated gassing test of powder in KOH would not allow to predict cell gassing.

The anode gassing rate of fine zinc powder is greater with lead-free zinc powder: 95% higher average gassing rate, 69% higher maximum gassing rate. Zn-In alloy powders showed moderate increases: similar average gassing rate, 36% higher maximum gassing rate. Leaded zinc powder with more fine particles showed 14% higher average gassing rate and similar maximum gassing rate. All comparisons are done with control zinc powder grade (500 ppm Pb, low fine fraction). Gassing was lowest for the Zn-Pb-In alloy with low fine fraction, but discharge performance of this powder was lower.

Gassing data were used to compute average and highest pressures in the tested RAM AA cells. The average pressure is calculated based on approx. 50 cells tested on the various gassing tests, and maximum pressure is the highest single value measured in any of those 50 cells. Data show that all zinc powders fall in an acceptable pressure range. Average pressure falls between 18 to 91 psi and maximum pressure was measured in the range of 99 to 199 psi depending on the zinc powder.

Discharge performance of lead-free zinc powders was best for powders with more fines and amounted to 92% of controls with 500 ppm Pb. The somewhat lower performance of lead-free zinc powders was caused by a lower operating voltage during discharge. Figure 4 compares the first discharge at a 300 mA constant current load of RAM AA cells containing lead-free zinc-indium alloy to the control zinc powder with 500 ppm lead. It can be seen that the lead-free zinc-indium alloy is running at a slightly lower operating voltage and polarizes somewhat earlier compared to leaded zinc powder. Figure 5 shows the same two zinc powders after 4 recharges on the fifth discharge. The two zinc powders perform close during 2/3 of the discharge, but then the lead-free zinc-indium alloy polarizes again earlier. In out-of-cell test, the lower performance of lead-free zinc powders was not really caused by a lower operating voltage, since the capacity was largely determined by a premature zinc polarization that induces a final voltage drop occurring at higher voltage with lead-free zinc powders (Fig. 3). This suggests that lead-free powders are more sensitive to zinc passivation.

Table 4 gives a comparison of capacity retention data between the different zinc powders in RAM AA cells for a simulated Motor & Toy test (discharge through 3.9 ohm to a cut-off voltage of 0.9V at room temperature). Fresh and cycled charged cells were stored for 2 weeks at 65°C. Overall capacity retention is best with zinc-lead alloys. Zinc-lead alloy with increased fine fraction showed a capacity retention of 90% for fresh cells and cycled charged cells. The pure lead-free zinc powder with low fine fraction showed a very poor capacity retention of 58% when stored as fresh cells, but had good capacity retention in the cycled charged state. By increasing the fine fraction of the pure lead-free zinc powder it was possible to achieve an excellent capacity retention of 96% in the undischarged (fresh) state. However, the capacity retention in the cycled charged state decreased to 73% compared to 83% for the Zn-Pb control group. The lead-free zinc-indium alloy with low fine content showed also a poor capacity retention in the undischarged (fresh) state, but had good capacity retention at a 81% level in the cycled charged state. Increasing the fine fraction of the lead-free zinc-indium alloy resulted in an acceptable capacity retention of 83% in the undischarged (fresh) state and 78% in the cycled charged state. It should be noted that the more severe Motor & Toy test was chosen for the capacity retention evaluation to see the effects of alloying pure zinc powder more clearly. Usually the capacity retention after high temperature storage would be measured at a low rate test (e.g. 43 ohm to 0.9V).

Table 5 presents a comparison of basic electrical parameters (OCV, CCV[1A@200ms]) and capacity after the drop test of RAM AA cells with the different zinc powders. It can be seen that the open circuit voltage (OCV) is hardly affected by the severe triple drop independent from the zinc powder used. In contrast, the closed circuit voltage (CCV) on the 1 ampere load is strongly influenced by the fine fraction in the zinc powder. The effect is more pronounced for lead-free zinc powders. Pure lead-free zinc powder showed a 9% drop in CCV voltage immediately after the drop and the lead-free zinc-indium alloy dropped even 19%. However, lead-free pure zinc powder and zinc-indium alloy showed an acceptable 3% drop in CCV voltage when the fine fraction was increased. Electrical performance on the Motor & Toy test after the drop showed a similar tendency. Lead-free zinc-indium alloy with low fine fraction showed poor capacity after drop at a 86% level. Zinc-lead alloy and lead-free pure zinc powder with low fine fraction were measured at a 3 to 4% lower capacity after the triple drop, which is acceptable. All zinc powders with more fines performed satisfactorily, even a slight increase in capacity was observed.

CONCLUSION

A series of zinc powders, containing lead and/or indium, were used in mercury-free Rechargeable Alkaline Manganese (RAM) cells. The presence of more fine zinc powder particles in the range of 44 to 75 microns improves the discharge performance by 10% for Zn-Pb powders and 5% for lead-free zinc powders. It was demonstrated that a lead-free Zn-In alloy powder can be used in mercury-free RAM cells that show acceptable gassing and a discharge performance of 92% compared to control cells with Zn-Pb powders. Mercury- and lead-free zinc powders with the fine particle fraction perform after severe mechanical impact (shock) with only minimal drop in closed circuit voltage at an 1 ampere load and no capacity loss.

REFERENCES

1. F.R. McLarnon, E.J. Cairns, *J. Electrochemical Soc.* **138**, 645 (1991).
2. K. Kordesch et al., *J. of Power Sources.* **51**, 61 (1994).
3. J. Daniel-Ivad et al., *Meet. Electrochemical Soc., New Orleans, LA, USA,* Abstr. No. 66 (1993).
4. J.-Y. Huot, *Int. Power Sources Symp.* 177 (1993)
5. German Patent 1,086,309 (1954) and Japanese Patent 33-3204 (1958)
6. K. Tomantschger, R. J. Book., J. Daniel-Ivad US Patent 5,424,145 (1995).
7. K. Tomantschger, E. Oran, K. Kordesch US Patent 5,162,169 (1992).
8. J. Daniel-Ivad, R. J. Book., K. Tomantschger, US Patent Pending (1994).

Table 1: Performance of RAM AA cells on IEC application tests. A comparison with Single Use Alkaline cells on Deep and Endurance Discharge is presented. Note: IEC stands for International Electrotechnical Commission.

MOTOR & TOY TEST

Use Condition		RAM		Single Use Alkaline		No. of Single Use Alkalines replaced by 1 RAM cell
	Cycle	Cumulative Service		Service		
		Ah	hours	Ah	hours	
Continuous Deep Discharge 3.9 Ω to 0.8 V	1	1.29	4.85	1.63	6.08	
	25	17.9	66.2			11
	50	28.0	103.5			17
Endurance Discharge 1h@3.9 Ω, then recharge, repeat until 0.8 V cut-off is reached		Cumulative Service Time, hours				
	1	1		1		
	7	7		7		
	300	300				43
		Single Use Alkalines had a rest period instead of recharge (similar to IEC 1h/day use pattern)				

PERSONAL CASSETTE PLAYER (WALKMAN®) TEST

Use Condition		RAM		Single Use Alkaline		No. of Single Use Alkalines replaced by 1 RAM cell
	Cycle	Cumulative Service		Service		
		Ah	hours	Ah	hours	
Continuous Deep Discharge 10 Ω to 0.9 V	1	1.51	13.2	1.96	17.0	
	25	21.5	186.3			11
	50	33.2	286.2			17
Endurance Discharge 4h@10 Ω, then recharge, repeat until 0.9 V cut-off is reached		Cumulative Service Time, hours				
	1	4		4		
	5	20		17.1		
	220	880				51
		Single Use Alkalines had a rest period instead of recharge (similar to 4h/day use pattern)				

TRANSISTOR RADIO TEST

Use Condition		RAM		Single Use Alkaline		No. of Single Use Alkalines replaced by 1 RAM cell
	Cycle	Cumulative Service		Service		
		Ah	hours	Ah	hours	
Continuous Deep Discharge 43 Ω to 0.9 V	1	1.92	68.9	2.29	80.1	
	25	25.3	885.1			11

PULSE (PHOTOFLASH/CAMERA) TEST

Use Condition		RAM	Single Use Alkaline	No. of Single Use Alkalines replaced by 1 RAM cell
	Cycle	Cumulative Service	Service	
		Pulses/Flashes	Pulses/Flashes	
Intermittent Deep Discharge 1.8 Ω, 15s/min to 0.9 V	1	417	531	
	25	5112		10

Note: Walkman is a registered Trademark of the Sony Corporation

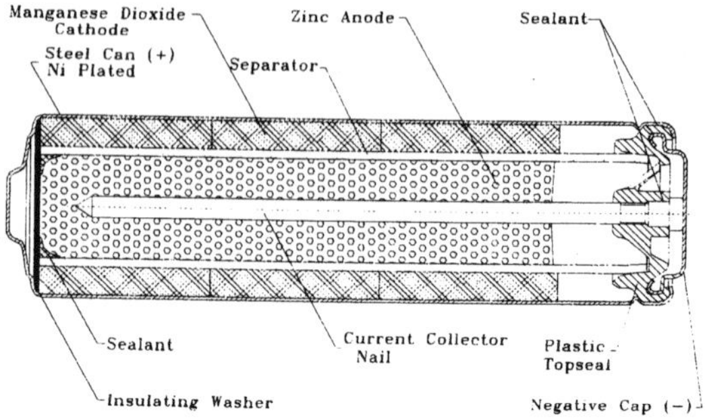

Figure 1: Cross sectional view of a RAM AA cell

Table 2: Properties of tested zinc powders.

ALLOY VARIABLES	ZINC POWDERS						
Lead, ppm	500		0		0		500
Indium, ppm	0		0		500		500
Particle Size Distribution							
Fine Fraction, -200 + 325 mesh	-	+	-	+	-	+	-
Mean Particle Size D50 (μm)	198	161	238	175	208	145	190
Width - particle size distribution	1.53	1.58	1.55	1.72	1.56	1.83	1.49
Density and Flowability							
Apparent Density (g/cc)	2.99	2.77	2.96	3.19	2.98	2.87	n/a
Tap Density (g/cc)	3.74	3.62	3.60	3.91	3.67	3.81	n/a
Flowability	50.1	51.6	n/a	n/a	n/a	48.3	n/a
Out-of-Cell Discharge @ 100mA/g Zn							
Capacity (mAh/g)	522	509	416	297	365	334	448
Slope of plateau (mAh/g-mV)	2.4	2.9	2.5	4.2	2.9	3.3	2.4
Out-of-Cell Gassing in 45wt.% KOH							
Zinc powder at 71°C @ 24h	-	-	++	++	++	++	+

In this Table, 0 refers to zero materials added. Actual analysis will show elements at low ppm levels.
n/a = not available

Figure 2: Particle size distribution of zinc powder alloy with 500 ppm lead.

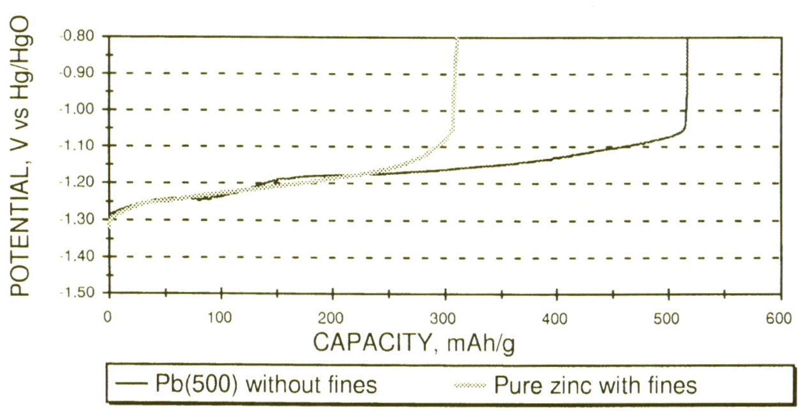

Figure 3: Comparison of lead-free zinc powder and zinc-lead alloy at a constant current discharge rate of 100 mA/g Zn in an electrochemical test cell setup with excess KOH.

Table 3: Comparison of electrical performance and gassing of RAM AA cells for a simulated Motor & Toy test (discharge through 3.9 ohm to 0.9 volt cut-off at room temperature) Gas volumes were collected after storage for 2 weeks at 65°C as fresh and cycled charged cells.

ALLOY VARIABLES	ZINC POWDERS						
Lead, ppm	500		0		0		500
Indium, ppm	0		0		500		500
Fine Fraction, -200 + 325 mesh	-	+	-	+	-	+	-
Indium added to zinc surface	yes	yes	yes	yes	yes	yes	yes
Average Capacity, Ah	**1.16**	**1.27**	**1.03**	**1.08**	**0.99**	**1.06**	**0.98**
Relative Performance %	100%	110%	89%	93%	86%	92%	85%
Relative Cumulative Performance %	100%	100%	89%	92%	91%	93%	94%
Average Gassing, mL	**1.2**	**1.4**	**1.0**	**2.3**	**0.9**	**1.2**	**0.5**
Average Pressure, psi	**47**	**54**	**40**	**91**	**36**	**48**	**18**
Relative Average Gassing %	100%	114%	85%	195%	76%	102%	38%
Max. Gassing, mL	**3.0**	**2.9**	**2.8**	**5.0**	**4.0**	**4.0**	**2.5**
Max. Pressure, psi	**118**	**113**	**111**	**199**	**159**	**160**	**99**
Relative Max. Gassing %	100%	96%	95%	169%	135%	136%	84%

In this Table, 0 refers to zero materials added. Actual analysis will show elements at low ppm levels.
n/a = not available

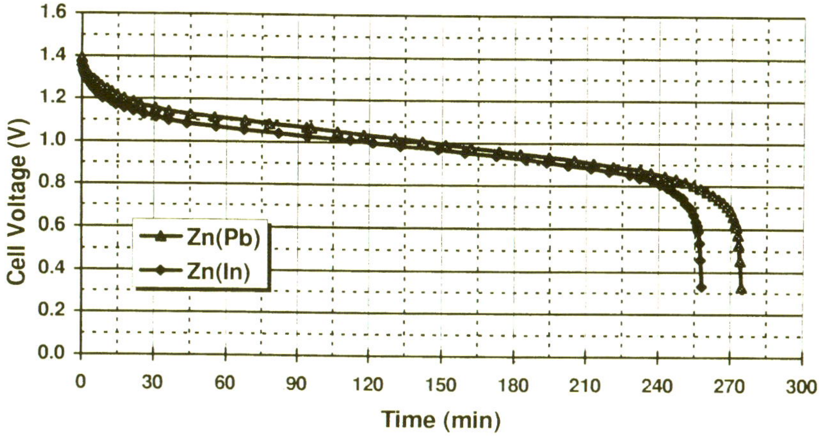

Figure 4: First discharge voltage curves of RAM AA cells at a constant current rate of 300 mA. Lead-free zinc-indium powder is compared to zinc-lead powder control.

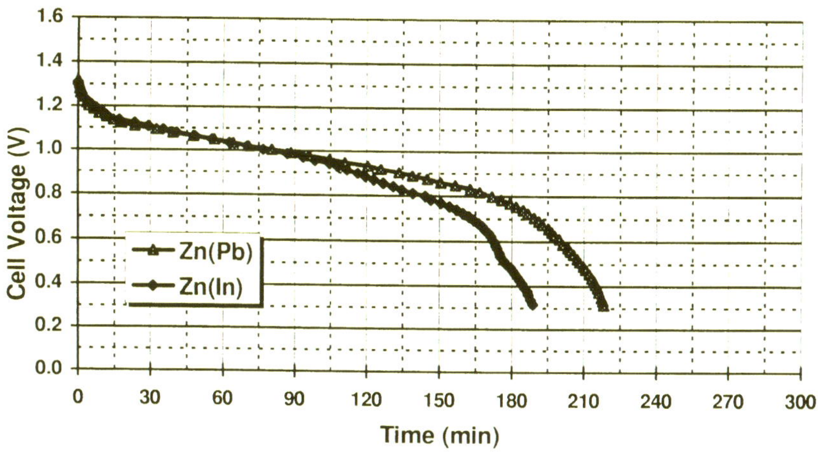

Figure 5: Fifth discharge voltage curves of RAM AA cells at a constant current rate of 300 mA. Lead-free zinc-indium powder is compared to zinc-lead powder control.

Table 4: Comparison of capacity retention of RAM AA cells with various zinc powders for a simulated Motor & Toy test (discharge through 3.9 ohm to 0.9 volt cut-off at room temperature) Fresh and cycled charged cells were stored for 2 weeks at 65°C.

ALLOY VARIABLES	ZINC POWDERS						
Lead, ppm	500		0		0		500
Indium, ppm	0		0		500		500
Fine Fraction, -200 + 325 mesh	-	+	-	+	-	+	-
Indium added to zinc surface	yes	yes	yes	yes	yes	yes	yes
Average Capacity, Ah	**1.19**	**1.30**	**1.18**	**1.20**	**1.08**	**1.13**	**1.08**
Relative Performance %	100%	109%	99%	101%	91%	95%	91%
Fresh Cells, cyc1 Ah	**1.19**	**1.30**	**1.18**	**1.20**	**1.08**	**1.13**	**1.08**
2wks65°C->cyc1 Ah	**0.96**	**1.19**	**0.69**	**1.15**	**0.73**	**0.94**	**0.80**
Capacity Retention %	80%	91%	58%	96%	68%	83%	74%
Cycled Charged Cells, cyc3 Ah	**0.90**	**0.93**	**0.77**	**0.89**	**0.81**	**0.84**	**0.74**
2cyc->2wks65°C->cyc3 Ah	**0.74**	**0.83**	**0.71**	**0.65**	**0.66**	**0.66**	**0.72**
Capacity Retention %	83%	90%	92%	73%	81%	78%	97%

In this Table, 0 refers to zero materials added. Actual analysis will show elements at low ppm levels.
n/a = not available, cyc = discharge/charge cycle

Table 5: Comparison of Basic Electrical Parameters (OCV = Open Circuit Voltage and CCV = Closed Circuit Voltage [1A@200ms]) and Capacity after drop test of RAM AA cells with various zinc powders for a simulated Motor & Toy test (discharge through 3.9 ohm to 0.8 volt cut-off at room temperature). Fresh cells were dropped from a height of 1 meter onto a concrete base as follows: First drop on positive end, then drop on negative end then drop on side with the same cell.

ALLOY VARIABLES	ZINC POWDERS						
Lead, ppm	500		0		0		500
Indium, ppm	0		0		500		500
Fine Fraction, -200 + 325 mesh	-	+	-	+	-	+	-
Indium added to zinc surface	yes	yes	yes	yes	yes	yes	yes
Average Capacity, Ah	**1.19**	**1.30**	**1.18**	**1.20**	**1.08**	**1.13**	**1.08**
Relative Performance %	125%	136%	123%	125%	113%	118%	113%
Drop Test (Triple drop)							
OCV before drop, V	1.55	1.56	1.57	1.57	1.57	1.57	1.57
CCV (1A@200ms) before drop, V	1.34	1.34	1.32	1.32	1.32	1.34	1.35
OCV after drop, V	1.55	1.56	1.57	1.57	1.55	1.57	1.57
CCV (1A@200ms) after drop, V	1.28	1.33	1.20	1.28	1.07	1.30	1.24
OCV after vs. before drop, %	**100%**	**100%**	**100%**	**100%**	**99%**	**100%**	**100%**
CCV after vs. before drop, %	**96%**	**99%**	**91%**	**97%**	**81%**	**97%**	**92%**
Capacity after drop, Ah	1.16	1.34	1.13	1.26	0.93	1.19	1.22
Capacity after drop, % of no drop	**97%**	**103%**	**96%**	**105%**	**86%**	**105%**	**113%**

In this Table, 0 refers to zero materials added. Actual analysis will show elements at low ppm levels.
n/a = not available, cyc = discharge/charge cycle

BONDED RECHARGEABLE ZINC ANODE

M.Weissenbacher, J.O.Besenhard, K.Kordesch

Institute of Inorganic Chemical Technology
Technical University of Graz
Stremayrgasse 16/III, 8010 Graz, Austria

Plastic bonded zinc electrodes were produced from metallic zinc and a plastic binder. For thicker electrode designs, a pore former was used in addition. These anodes were predominantly tested in rechargeable alkaline manganese dioxide - zinc cells, but the technology is pricipially adaptable to all other alkaline battery systems that use zinc anodes. Dimensionally stable zinc anodes in cylindrical cell designs outperform gel electrodes with respect to higher conductivity, an improved resistance against shock impact, and far better gas recombination properties. Both, conductivity and mechanical stability, could be additionally improved by sintering the anodes with low melting point metals. MnO_2/Zn button cells and stack-type 9V cells have been built.

BACKGROUND OF THE WORK

RAM (Rechargeable Alkaline Manganese Dioxide) cells, developed in Graz with production methods scaled up by BTI (1), were licensed to Rayovac® and successfully introduced into the US market in fall 1993 due to some outstanding features such as good shelf life, low cost and environmental friendlyness. However, this battery system using gelled zinc powder anodes is still far from being perfect, as there is a significant capacity fade in the first cycles, fortunately flattening to a plateau of constant discharge capacity after about 15 cycles. Furthermore, the discharge capacity strongly depends on the load.

The identification of limiting factors in complete RAM cells is complicated by the fact that their overall behaviour is determined by a complex combination of single component properties. Nevertheless, some problems can be easily attributed to the gelled anodes (in particular loss of active mass due to corrosion and poor contact to the current collector causing poor results in drop tests) or to the microporous separator (increased internal resistance). The discharge capacity fade of RAM cells is usually allocated to the cathode as it is a typical manganese dioxide electrode feature, but reference electrode studies have also shown a strong decline in anode performance, in particular for applications which are usually operated down to low voltages or even breakdown, like flashlights or radios.

Gelled zinc anodes made from zinc powder dispersed in gelly electrolyte are commonly used in primary and secondary alkaline manganese dioxide cells. In RAM batteries several inadequate features of the zinc gel anode became evident:

- A strong capacity fading after deep discharge as it is illustrated by figure 1. By contrast, the anode capacity remains quite constant when RAM cells are discharged to 0.9V-0.8V only, which is the minimum working voltage for many applications.
- A strong dependence of the discharge capacity on the current density (2). As the cycle life of RAM cells strongly depends on the „switch off" by the capacity limiting zinc anode (to avoid deep discharge of the manganese dioxide electrode) it is somewhat difficult to balance RAM cells perfectly for all applications. The ideal zinc anode should passivate just before overdischarge of MnO_2 occurs at any load. Moreover, the potential of the zinc anode should be constant during discharge, followed by a sudden rise to stop the discharge of the cell.
- A bad contact situation. After mercury has been replaced by other corrosion inhibitors, e.g. polymers that partly cover the zinc surface, it became obvious that mercury has also had a beneficial effect on the contact situation by providing „bridges" between the zinc particles and also between the zinc particles and the current collector. Improving the contact situation by increasing the zinc particle density is not very helpful in zinc limited RAM cells. As the utilization of the manganese dioxide cathode decreases strongly with increasing thickness (3,4), the cathode layer has to be kept as thin as possible in most cell designs. Because of the required zinc limitation, standard cell designs are not really optimum for sleeve-type RAM cells. In particular large cell designs like C and D are less suited than slim ones like AA and AAA.

A reliable contact situation is not only important for specific, e.g. military, applications, but is also of big relevance for standard consumer cells considering that a big portion of the sales falls to toys. The quality of the contact situation is usually checked by drop tests followed by a measurement of the cell's inner resistance or by vibration tests. These tests show that the contact situation in mercury free gels is much worse than in anodes which use amalgamated zinc (figure 2) (5). The drop test results reflect a very loose structure of the zinc gel anode. For this reason, we have started a program to develop a plastic bonded zinc anode with a dimensionally stable zinc structure. High porosity, shaped zinc electrodes showing a high current output and a good cycling efficiency could not only be used for RAM cells, but also for other alkaline battery systems which start off using metallic zinc, like silver-zinc, mercury-zinc, and zinc-air.

Considering the future RAM cell development, the construction of plastic bonded electrodes is essential. To extend the application range of RAM cells to devices like emergency lights, electrical shavers, telephones and others of higher voltage requirements, the overdischarge and overcharge behaviour of cells in battery packs of up to 9V has to be improved. The overcharge properties strongly depend on the anode, as the oxygen gas that is generated in the cathode on overcharge has to be consumed in the anode by the reaction of zinc to zinc oxide (6). Access of the oxygen gas to the anode strongly depends on the hydrophobicity of the anode, which can be controlled by the amount of plastic in the electrode (7). To achieve competitive RAM battery products for applications like camcorders, computers, etc., the development of wound, plate, and other power cell designs is necessary. These cell types need thin plastic bonded electrodes.

Another advantage of dimensionally stable anodes is the possibility to apply a separator coating to the anode by a simple dipping method. RAM cells currently use a two layered separator, one layer of cellulose, which stores electrolyte, and one microporous layer of cellophane (regenerated cellulose), which prevents the penetration of the separator by zinc dendrites during charging. There are several coating formulations which could possibly replace the cellophane layer, but coating the inner side of the cylindrical cathode pellets (e.g. by rotating) is hardly practicable in a production line and coating of the gel anode is impossible. Forst and Vielstich avoided the formation of dendrites by coating the surface of a zinc anode with a dispersion of cureable colloid particles, like styrol-acrylat copolymers, which are resistent against KOH (8).

PRINCIPAL PROCEDURES

There are two main routes to produce plastic bonded electrodes: the dry mix and the wet mix fabrication (figure 3). The wet mix fabrication guarantees a homogeneous distribution of the polymer that is dissolved or dispersed in some liquid. This method is well suited for the production of very thin electrodes, which do not necessarily require the use of additional pore formers.

The dry mix fabrication is based on a good mechanical mixing, followed by the pressing into the right shape. It is suited for thicker electrodes requiring the use of pore formers. Our 'dry mix' plastic bonded zinc electrodes were generally made from metallic zinc powder, a plastic binder, a pore former, and zinc oxide, which acts as an overcharge protection. The use of other additives, which influence the porosity and the zinc(ate) precipitation is optional. In principle soluble or volatile filler materials may be chosen as pore formers.

FILLER MATERIALS

Soluble Fillers

Many different salts are suited as soluble filler materials. A procedure for the production of high porosity electrodes using a soluble filler is given e.g. by Morrel and Smith [9] and by Sunu and Bennion (10). The authors used a mixture of NaCl powder and Zn(Hg) powder pressed onto a backing plate. The resulting electrode was soaked in distilled water for 24 hours to give a highly porous structure with a porosity of 0.6. We have used NaCl and $Na_2SO_4 \cdot 10\ H_2O$ as soluble fillers and obtained electrodes according to the descriptions in reference (10). The removal of the soluble filler material is generally milder than a heat treatment, however, it is too time consuming for a technical process. Furthermore, the selective extraction of a soluble filler material may be a problem, as the organic corrosion inhibitor, which forms a layer on the zinc surface, has to be insoluble in the respective solvent. If polyethyleneglycol (PEG) is used as corrosion inhibitor, water cannot be used to remove the pore maker, as PEG is soluble in water.

Another possible ´soluble´ filler material is dry potassium hydroxide, which can be activated with water. These electrodes are reported to perform equivalent to anodes which use aqueous KOH, but have to be prepared from specially dried zinc powder under controlled low humidity conditions (11). Furthermore the handling of the electrode until the cell assembly requires exclusion of carbon dioxide to avoid carbonate formation.

Volatile Filler

Volatile filler materials should decompose at temperatures as low as possible, as long as they are stable at the electrode manufacturing temperature and pressure. Common volatile pore formers are the ammonium salts of carbonic acid or other weak organic acids, such as formic, acetic or oxalic or these organic acids themselves (12). All decomposition products should be chemically inert and volatile as well. For this reason we favour ammonium carbamate ($NH_4NH_2CO_2$) over ammonium carbonate ($(NH_4)_2CO_3$) or ammonium hydrogencarbonate (NH_4HCO_3) as no water is produced in the decomposition of the carbamate (might cause zinc corrosion).

Table 1: Informations about some commonly used volatile pore formers (13).

Formula	Name	Thermal Decomposition Reaction	Temp. [°C]	Densities [g/ml]*	Solubilities
$NaHCO_3$	*Sodium hydrogen carbonate*	$2\ NaHCO_3 \rightarrow Na_2CO_3 + H_2O + CO_2$	50+	2.159	100g/l H_2O......20°C
$(NH_4)_2CO_3$	*Ammonium carbonate*	$(NH_4)_2CO_3 \rightarrow 2\ NH_3 + H_2O + CO_2$	*58*		1g/ml H_2O .....15°C insol.: alcohol
NH_4HCO_3	*Ammonium hydrogen carbonate*	$NH_4HCO_3 \rightarrow NH_3 + H_2O + CO_2$	107.5	1.58	11.9g/100ml....0°C dec. in hot water insol:alcohol,acetone
$NH_4NH_2CO_2$	*Ammonium carbamate*	$NH_4NH_2CO_2 \rightarrow 2\ NH_3 + CO_2$	60	0.750**	very sol.:cold water dec. in hot water very sol.:NH_4OH slightly sol:alcohol insoluble: acetone

*) Densities important to calculate porosity! **) bulk density

The hygroscopic properties of the filler may be critical in the manufacturing process. Ammonium carbamate is less hygroscopic than ammonium hydrogencarbonate.

Certain acidic fillers (NH_4Cl, NH_4Br, NH_4I) serve the dual role of i) reactant and ii) pore maker. Their fluxes remove zinc oxide from the zinc surface and provide preliminary connections between the cleaned zinc particles. To make use of this effect, the zinc/filler mixture is slowly heated up to max. 320°C under vacuum (reduced pressure). The reaction products are volatile under these conditions (14,15). Equation [1] shows the reactions of NH_4Cl with ZnO.

$$ZnO + 2\,NH_4Cl \rightleftarrows \begin{cases} Zn(NH_3)_2Cl_2 + H_2O \\ \updownarrow \\ ZnCl_2 + 2\,NH_3 + H_2O \end{cases} \qquad [1]$$

Special fluxes, e.g. NH_4Cl, $ZnCl_2$, $AlCl_3$, and hydrazin hydrochloride are reported to react with the zinc oxide cover under high pressures in exothermic reactions (16). The heat of reaction is sufficient to sinter the zinc particles. Reaction products and the flux are removed by a mixture of water and acetone.

In both cases, i.e. soluble or volatile fillers, the distribution of the pore sizes is determined by the particle size distribution of the filler material.

PLASTIC BINDER

The binder should provide a stable network in which the zinc particles are distributed. It should be chemically inert and mechanically stable under the required conditions (pH and temperature). On the other hand, the binder should have some affinity to the electrolyte. If the plastic is too hydrophobic, the KOH impregnation could be insufficient or too slow. The hydrophobicity of the electrode also affects the gas recombination properties (7). In RAM cells the oxygen gas, which is produced in the cathode on overcharge, can be recombined by the reaction with zinc (to give zinc oxide), so long as sufficient hydrophobic reaction sites are available. The gas release properties, which are important to release hydrogen produced in the anode to be recombined in the cathode, are also depending on hydrophobicity. Swelling of the plastic may cause a positive effect as long as the structure remains stable, as more electrolyte can be stored in the swollen electrode.

Looking for binders that could be suitable for RAM cell zinc anodes, all binders which are reported to be compatible with manganese dioxide in KOH can be considered. Poly(butadiene styrene), poly(ethylene), PTFE, poly(styrene), poly(vinylacetate), poly(vinylchloride), poly(vinylalcohol), polyolefines, partially fluorinated polyolefines, poly(isobutylene), methacrylate, polysulfone, epoxy have been used previously either as powder or in form of emulsions or solutions to give stable manganese dioxide cathodes (17).

Many different plastics have also been proposed as binder materials for zinc anodes made from zinc oxide and/or metallic zinc, most of them for the use in silver-zinc and nickel-zinc cells. Much attention has been paid to poly(tetrafluoro ethylene) (PTFE) (7, 18-22), but zinc anodes made from other polymeric binders such as poly(vinyl alcohol) (23) poly(styrene), poly(methyl methacrylate), poly(vinyl chloride), poly(carbonate), poly(vinyl acetate), poly(isoprene), poly(ethylene) and poly(acrylonitrile styrene) have also been studied intensively (24). It was reported that poly(styrene) and poly(vinyl chloride) reduce both the proportion of accessible zinc in the electrode and the cycle life, while poly(methyl methacrylate) and poly(vinyl acetate) significantly increase the cycle life at low levels of addition. Poly(carbonate) is reported to permit total zinc utilization, but also to greatly increase cycle life by stabilizing the zinc electrode structure against shedding (24).

Many battery manufacturers use solutions or suspensions of polymers, as a good mixing is sometimes difficult when a powdery binder is applied, due to the big differences in specific weights of the components. A procedure to fabricate a porous zinc electrode for alkaline manganese dioxide cells is given by T.Chobanov (15). The elastic anode is made from zinc powder, ammonium hydrogencarbonate and poly(isobutylene) (BASF Oppanol B 200) dissolved in cyclohexane. After intensive mixing the solvent is removed by filtering through a glass frit. This results in a pasty mass that can be extruded. The filler is then removed by a heat treatment.

Despite the fact that numerous plastic binders for electrodes have already been described in the literature, there will be still many new ones, as the polymer science is rapidly growing, developing new (co-, block-) polymers. The overview of today`s variety of commercially availabel plastic materials is faciliated by reference books like the „Materials Selector" (25).

METAL AND METAL OXIDE ADDITIVES

As mentioned earlier, the reduction of the mercury content in zinc anodes down to 0% mercury addition severely decreased the contact situation in zinc gels. It has been shown that the addition of certain metals -or their respective oxides- can considerably improve the contact situation (5). Figure 2 shows the increase of the short circuit current of AA RAM cells, when one percent of SnO is mixed to the mercury free zinc gel. The tin oxide can also be dissolved in the potassium hydroxide electrolyte. The addition of metal oxides (and metals) to zinc anodes is well established in the literature, predominantly aiming at the replacement of HgO by other additives, mainly heavy metal oxides which suppress the hydrogen evolution and do not accelerate shape change. Lead oxide (PbO) and cadmium oxide (CdO) are reported to significantly reduce the rate of shape change, while thallium oxide (Tl_2O_3) minimizes the hydrogen evolution rate (26). Exellent results were obtained with small additions of binary mixtures of the oxides or hydroxides of high hydrogen overvoltage metals to PTFE bonded zinc oxide anodes (27). Typical combinations are 1% of PbO or SnO_2 with 0.5% to 2% Tl_2O_3, CdO, $In(OH)_3$, or Ga_2O_3. The electrochemistry of the mentioned metal oxide additives in pasted zinc anodes has been extensively studied (28).

Metal, -oxide, -hydroxide, -salt additives mentioned in the patent literature are lead, cadmium and tin salts (29-34), cadmium metal (31), cobald hydroxide (35), tin metal (36), inorganic titanates (37), calcium oxide (38) and bismuth oxide (38). Bi_2O_3, CdO, and acetylene black, also improved the zinc anode performance most in a test series for zinc plates in nickel/zinc cells. In these tests a large number of different additives has been screened to see, if they improve zinc utilization over 100 deep discharge cycles of the zinc electrode. The beneficial effect of bismuth appears to be due to the conductive matrix throughout the zinc anode consisting of needle-like deposits from reduced Bi_2O_3 (39).

All the above-mentioned metal oxides, which are reported to have beneficial effects on zinc anodes, are reduced in the zinc electrode due to their more noble character. For this reason it is also possible to use directly the corresponding metals. Sintering of zinc anodes with low melting point metals and alloys provides strong intermetallic connections between the zinc particles. This should considerably improve i) the contact situation throughout the anode and ii) the mechanical stability of the electrode. As our porous electrodes have to be heated up to remove the volatile filler, the sintering is not an additional production step. However, as the anode has to be heated to significantly higher temperatures, this has to be done in an oxygen-free atmosphere. The respective metal/alloy should not increase the corrosion rate and not be environmentally harmful like e.g. lead, cadmium, or thallium. Besides tin and bismuth, other metals were considered that are known to be used as additives in battery grade zinc alloys. Some of these metals are added to zinc to increase the hydrogen overvoltage on the surface of zinc powder (indium, bismuth, lead, calcium), some to smooth the zinc surface (aluminium, calcium, barium) (40). Table 2 contains the melting points of selected metals and alloys, many of which are too expensive to be used in higher percentages. The eutectics of tin and zinc (92/8, mp.199°C) and of bismuth and tin (58/42, mp.138.5°C) have low melting points and might be affordable (table 2).

Table 2: Melting points of Zn, Sn, Bi, In and eutectics (41)
(wt% of the component in the first column\melting point in °C)

	+ zinc	+ tin	+ bismuth	+indium
zinc	100 | 419.4°C			
tin	92 | 199°C	100 | 232°C		
bismuth	97 | 254°C	60 | 140°C	100 | 271°C	
indium	96 | 143°C	47 | 117°C	35|110°C, 52|89°C	100 | 157°C

RESULTS AND DISCUSSION

Thin (<1mm) and thick, but highly porous electrodes have been prepared from 0% mercury addition zinc. Thin electrodes were produced „flat square" and „round" according to figure 3 (wet procedure) and tested in manganese dioxide - zinc bipolar cells.
Thicker anodes (>1mm) were produced by the pressing method and the use of filler materials. This way discs and cylindrical electrodes have been made. Disc anodes (∅:2cm) have been tested in rechargeable alkaline manganese dioxide button cells and stack-type 9 V batteries (Figure 4). Cylindrical anodes were produced in AA size using a pressing cylinder and a split part, rendering it possible to press the anode directly on to the current collector. The current collector is the standard brass nail fixed to the top of the battery, serving as the negative terminal of the cell and including the plastic sealing facilities. Figure 5 shows a photograph of such a AA anode with the brass nail. AA anodes were made according to the flow sheet diagram 'dry mix' in Figure 3. As mentioned above, we prefer the use of volatile fillers, as the soluble filler procedure is too slow. Removal of volatile fillers in AA anodes is fairly quick, as the anodes can be placed on the top of the current collector and the heat is directly conducted into the center via the brass nail. Grillo zinc G40-O/300 Bi/300In was used for all anodes and polyethylenglykol (PEG) 600 was used as an organic corrosion inhibitor.

Considerations were also made about the possibility, to perform the temperature treatment in ordinary atmosphere and not in vacuum. Zinc melts at 419°C and evaporates at 900°C. Significant oxidation is mentioned to start at 225°C (42), but it depends strongly on the particle size (zinc particles in the micron and sub-micron range are pyrophoric when exposed to the air). Decomposition of PEG is not observed below 100°C. As mentioned above, several volatile fillers decompose far beyond 100 °C. Plastic binders need very different temperature ranges to be sintered, if they have to be sintered at all. Poly(tetrafluoro ethylene) (PTFE) bonded fuel cell electrodes are typically sintered at about 300°C for 20 minutes to improve their mechanical strenght (43,44). If not required for the plastic binder, the electrode has not necessarily to be heated up to more than 100°C, which could be just low enough to avoid zinc oxide formation. Results of electrodes prepared with PTFE, heated up to 90°C only, gave generally good results, although we sometimes obtained electrodes of lower specific capacity which was devoted to zinc oxide formation even at quite low temperatures.

To eveluate the cycling behaviour of the plastic bonded anodes, they were cycled versus oversized manganese dioxide cathodes, which were discharged much less than one third of the first electron capacity. Figure 6 shows the cycle life of a plastic bonded anode with a maximum specific capacity value of about 700 Ah/l at a current density of 110mA/cm^2 (geom) or ca. 260 mAh/g, which corresponds to an efficiency of about 46 %. Additives like Bi_2O_3 (5 wt%) or metallic tin (7 wt%) significantly increased the short circuit current (SCC) (Sn: 154%, Bi_2O_3: 131% compared to standard electrode without additive) and also the specific capacity at higher loads, which is equivalent to a higher zinc utilization.

Figure 7 demonstrates that the cycle life of RAM AA cells, which contain plastic bonded zinc anodes is equivalent to standard cells with zinc gel electrodes. The value of about 1000 Ah/l at a current density of 40 mA/cm^2 has to be reached by filling out the given AA anode volume space. Depending on the sintering conditions, cells with plastic bonded anodes show higher SCC and outperform standard cells in drop tests.

BET (3 point method/N_2) surface measurements resulted in typical values of about 3 m^2/g for anode material made with 30 wt% pore former. The maximum achieveable density of the original electrode body before the heat treatment is about 3.2 g/cm^3 corresponding to a porosity lower than 10%. The final porosity and pore size distribution were detected by the mercury intrusion method and depend on the particle size of the zinc as well as on the filler material and also slightly on the applied pressure. A porosity of about 65% was adjusted for AA cells (Figure 8).

The possibility to overcharge RAM cells applying constant current depends on i) the amount of ZnO in the anode, ii) the current density iii) the zinc particle size, and iv) the porosity and hydrophobicity of the anode, which has to accept the oxygen gas coming from the cathode. The possible constant current to overcharge AA RAM cells could be increased to 75 mA by the use of plastic bonded anodes (Figure 9).

ACKNOWLEGMENT

The authors are grateful to BTI for the financial support and the supply of AA cathodes and cell parts and would like to thank to Grillo-Werke Aktiengesellschaft for providing zinc samples.

REFERENCES

(1) Battery Technologies Inc., 30 Pollard Street, Richmond Hill, Ontario, Can L4B 1A8
(2) P.G.Cheeseman, M.G.Stock, Power Sources 10, 217 (1984)
(3) W.J.Wruck, Doctoral Dissertation, University of Wisconsin, Madison (1984)
(4) Ch.Faistauer, Doctoral Dissertation, Technical University of Graz, Austria (1993)
(5) M.Weissenbacher, Diploma Thesis, Technical University of Graz, Austria (1993)
(6) K.Kordesch, M.Weissenbacher, P.Urdl, Power Sources 15, 466 (1995)
(7) F.Solomon, Pat.US 3,536,537 (1970)
(8) H.Forst, W.Vielstich, German Pat. DT 23 60 026 A1 (1973)

(9) D.H.Morrel, D.W.Smith, 5th Internat.Power Sources Symposium, 207 (1967)
(10) W.G.Sunu, D.N.Bennion, J.Electrochem..Soc., 127 (9), 2017-25 (1980)
(11) A.Charkey, R.Di Pasquale, Proc.Power Sources Conf. 22, 117 (1968)]
(12) L.I.Redey, D.R.Vissers, J.Prakash, Pat.US 5,283,135 (1994)
(13) Handbook of Chemistry and Physics, 57th Edition (1977)
(14) F.Przybyla, F.J.Kelly, Power Sources 2, 373-387 (1968)
(15) T.Chobanov, German Pat. DE 3416728 A1, (1984)
(16) P.R.Mallory & Co, DE 15 33 320 C3
(17) K.Kordesch, A.Kozawa, T.S.Lee, C.A.Roberts, UCC Res. Mem. CRM-300, (1973)
(18) G.Rampel, Pat.US 3,360,781 (1971)
(19) A.Charkey, Power Sources 4, 93 (1972)
(20) A.Charkey, 26th Power Sources Symposium, 87 (1974)
(21) M.Cenek, O.Kouril, J.Sandera, A.Touskova, M.Calabek, P. Sources 6, 215 (1977)
(22) Energy Research Corp., N.Y., Pat.GB 1 558 725 (1977)
(23) S.P.Poa, S.J.Lee, J.Appl. Electrochem. 9, 307 (1979)]
(24) N.A.Hampson, A.J.S.McNeil, J.Power Sources 15, 261, 254, 61 (1985)
(25) N.A.Waterman, M.F.Ashby; Elsevier, 1991
(26) O.Wagner, A.Himy, Proc.27th Power Sources Conf., 135 (1976)]
(27) A.Himy, O.C.Wagner, Pat.US 4,084,047
(28) J.McBreen, E.Gannon, Electrochimica Acta 26, 1439 (1981)
(29) Yardney Electric Corp., Pat.GB 2,006,512A (1977)
(30) Yardney Electric Corp., Pat.GB 2,007,420A (1977)
(31) Energy Research Corp., Pat.GB 1,558,725
(32) Yardney Electric Corp., Pat.US 4,118,551 (1977)
(33) US Secretary of the Navy, Pat.US 4,084,047 (1976)
(34) Energy Research Corp., Pat.US 4,022,953 (1975)
(35) Nat. Physical Lab.of Israel, Pat.US 4,152,224 (1976)
(36) Varta AG, German Pat.DE 2,748,992 (1977)
(37) Yardney Electric Corp., PatGB 1,533,698 (1976)
(38) Tokyo Shibaura Electric Co. Ltd., Pat.GB 1,530,961 (1975)
(39) J.McBreen, E.Gannon, J.of Power Sources 15, 169 (1985)
(40) H.Yoshizawa, A.Miura, Progress in Batteries & Mattery Material 12, 132 (1993)
(41) M.Hansen (1958), R.P.Elliott (1965), F.A.Shunk (1969), Constitution of Binary Alloys + Suppliment, McGraw-Hill Inc.
(42) Römpp, Georg Thieme Verlag (1991)
(43) K.Kordesch; Brennstoffbatterien, Springer, 1984
(44) M.Schautz, Doctoral Dissertation, Technical University of Graz, Austria (1985)
(45) E.Kahraman, L.Binder, K.Kordesch, J.Power Sources 36 (1991)

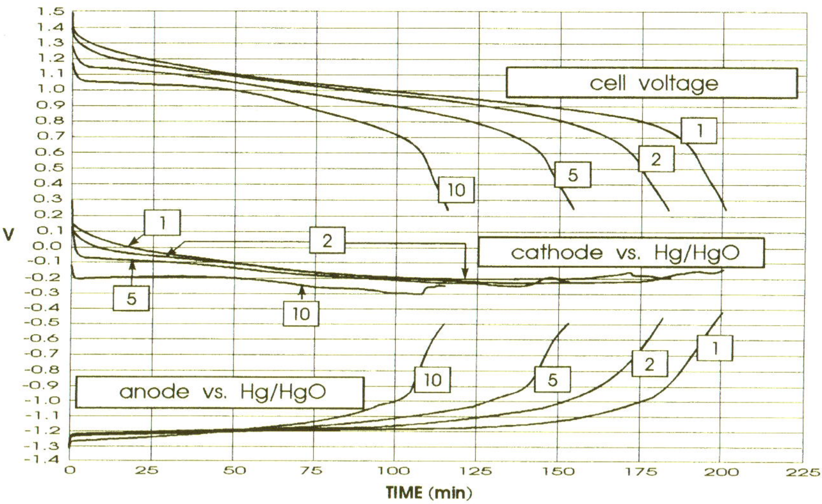

Figure 1: Reference electrode studies have shown a strong decline in anode performance, if the zinc gel anode is totally discharged (BTI-data, AA RAM, 300 mA disch. current).

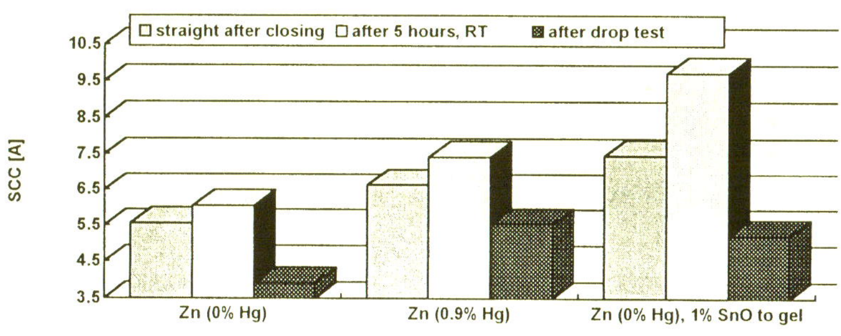

Figure 2: Short Circuit Current (SSC) of AA RAM cells with and without Hg, Sn (5)

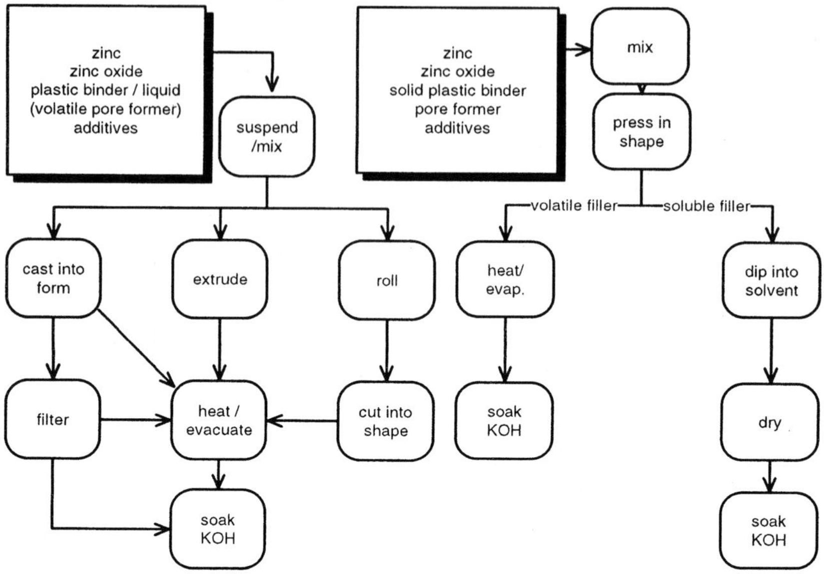

Figure 3: Flow-sheet diagrams for the production of electrodes according to the wet (left side) and the dry (right side) mix fabrication method.

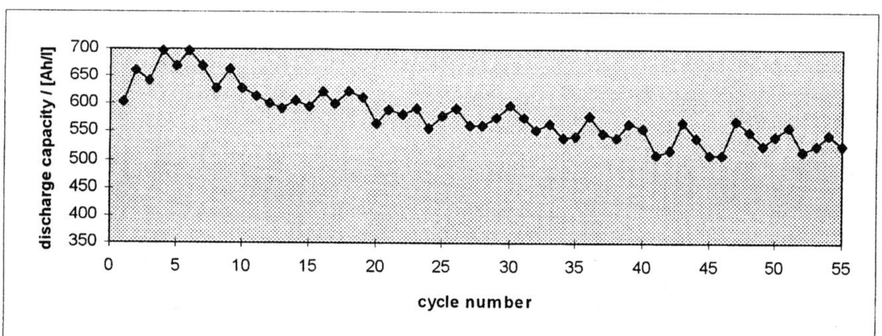

Figure 6: The cycle life of a plastic bonded anode with a maximum specific capacity value of about 700 Ah/l at a current density of 110mA/cm^2 (geom) or ca. 260 mAh/g, which corresponds to an efficiency of about 46 %.

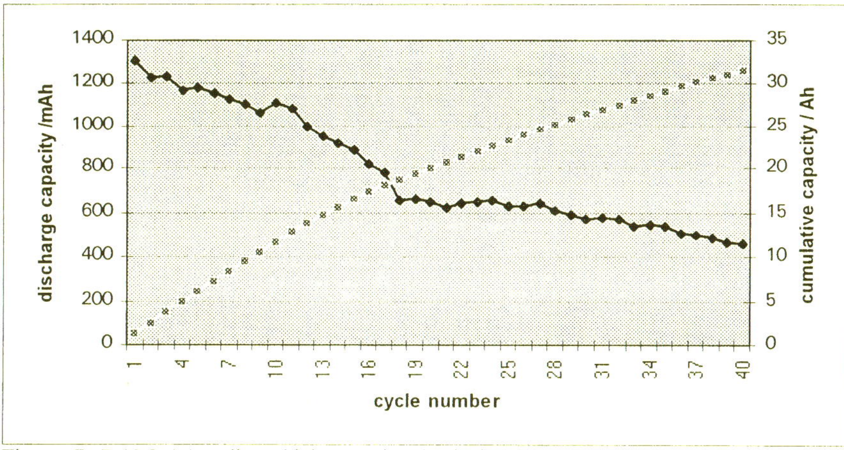

Figure 7: RAM AA cells, which contain plastic bonded zinc anodes, are equivalent to standard cells, which use a zinc gel, in regard to cycle life, but are superior in shock tests. Cycling conditions: load: 3.9 Ohm, COV: 0.9V, Charging voltage: 1.72V for 15 hours.

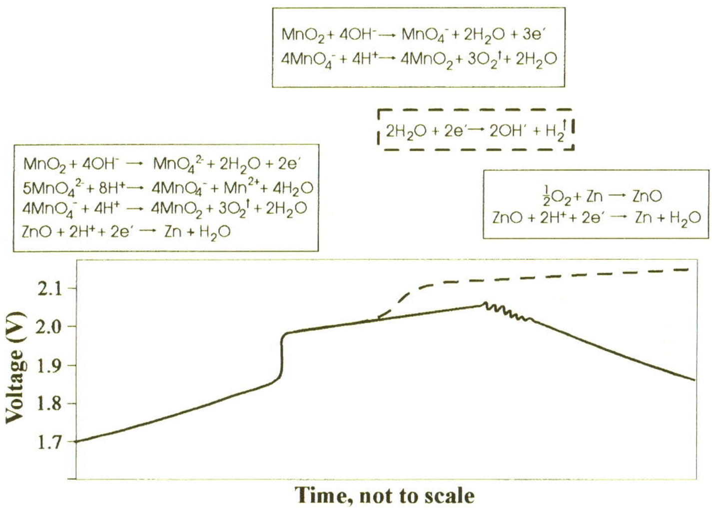

Figure 9: The possibility to overcharge RAM cells depends on the quick recombination of oxygen gas in the anode. The early development of O_2 in the cathode to avoid further oxidation of MnO_2 can be achieved by the use of certain catalysts (Co_3O_4, V_2O_5) (45).

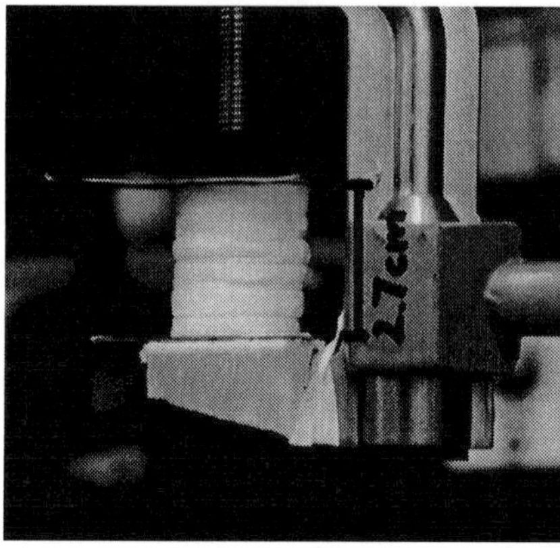

Figure 4: Stack-type 9V battery

Figure 5: Plastic bonded cylindrical AA zinc anode, directly pressed on to the current collector brass nail

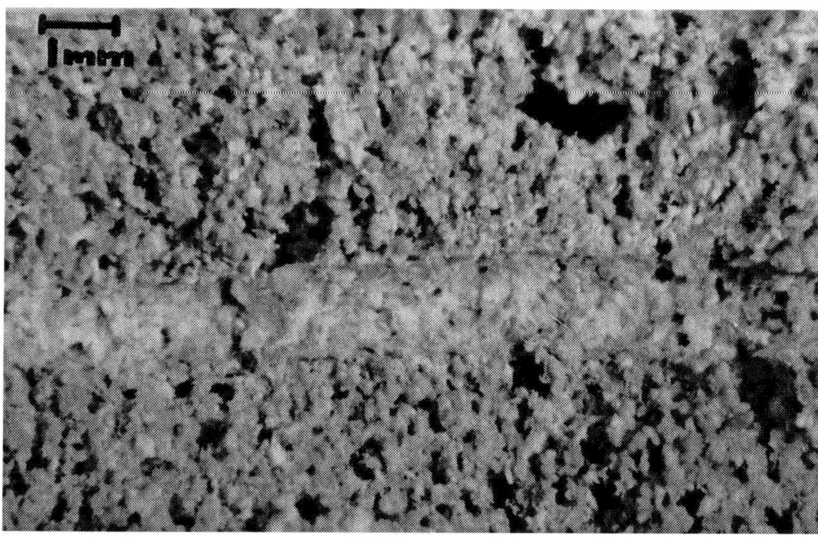

Figure 8: Photo of cylindrical anode with pores and print of nail

New Generation of Rechargeable Zinc-Air Batteries

S. Müller[1], F. Holzer[1], Chr. Schlatter[2],
C. Comninellis[2], and Otto Haas[1]

[1]Paul Scherrer Institut,
CH-5232 Villigen PSI, Switzerland

[2]Swiss Federal Institute of Technology,
CH-1015 Lausanne, Switzerland

For the development of a electrically rechargeable zinc/air battery, $La_{0.6}Ca_{0.4}CoO_3$-catalyzed (perovskite) bifunctional oxygen electrodes and pasted zinc electrodes were prepared and tested in a monopolar zinc/oxygen cell. In addition, a bipolar Zn/oxygen stack was tested using reticulated copper foam as substrate for the zinc deposit. The cells were cycled in moderately alkaline ZnO-saturated electrolytes with KF as an electrolyte additive. The maximum power as well as the cycle life of the cells was investigated. The microstructure and the performance of the electrodes before and after cycling were studied experimentally using physical (mercury porosimetry), electrochemical and imaging (X-ray photography) techniques.

INTRODUCTION

In metal/air batteries only the metallic active material has to be stored in the negative electrode, while oxygen is continuously provided to the positive electrode from the ambient air. For zinc/air cells the theoretical specific energy is 1085 Wh/kg, based on the molecular weight of ZnO (658 Ah/kg) and the theoretical cell voltage (1.65 V). Practical values of 90 to 300 Wh/kg for secondary and primary batteries, respectively, have already been demonstrated (1). The development of powerful and durable rechargeable zinc-air batteries will lead to cost-effective, high-specific-energy systems that are urgently needed for electric vehicles, stand-alone photovoltaic applications or as power sources for portable electric equipment.

The key task for the technical realization of electrically rechargeable zinc-air batteries is the development of an efficient, long-lived bifunctional oxygen diffusion electrode and of a zinc electrode with no significant shape change or dendrite formation. More recently, LBL (Lawrence Berkeley Laboratory) developed a pasted zinc electrode with a long cycle life (2), mainly for the

Zn/NiOOH battery. Shape change, which typically degrades Zn electrodes, was not the major problem in these cells, since they used moderately alkaline ZnO-saturated electrolyte and electrode or electrolyte additives ($Ca(OH)_2$ and K_2CO_3/KF, respectively), which lowers the solubility of the zinc electrode discharge products and inhibits redistribution of the Zn (2).

Significant progress has been made in recent years concerning the development of bifunctional catalysts. Shimizu et al. proposed a bifunctional oxygen electrode using $La_{0.6}Ca_{0.4}CoO_3$ as a catalyst (3). Together with LBL, we improved this electrode from the point of view of cycle life. The corrosion rate of the electrode material could be lowered considerably using graphitized carbon as a catalyst support material (4). In addition a two-layer bifunctional oxygen electrode was developed. Such porous two-layer perovskite-catalyzed diffusion electrodes, which were prepared by the rolling technique, exhibited constant potentials of O_2 evolution (2 V vs. Zn, 10 mA/cm^2) and reduction (1.2 V vs. Zn, 10 mA/cm^2) during 1300 h (130 cycles) (5).

In this contribution we present some results obtained with a monopolar 25 cm^2 Zn/oxygen cell and with a bipolar cell stack. In the bipolar battery, the Zn was homogeneously (6) deposited in a reticulated Cu foam. In order to avoid formation of soluble $Zn(OH)_4^{2-}$ during discharge, we worked with an insoluble discharge product. For this purpose a moderately alkaline electrolyte in which insoluble ZnO is formed during discharge was applied for the cycling test.

While in the monopolar cell configuration the zinc electrode is sandwiched between two bifunctional oxygen electrodes, in the bipolar cell arrangement only one cathode faces each anode.

EXPERIMENTAL

Monopolar Cell

In the monopolar cell, as depicted in Figure 1a, a pasted Zn/ZnO electrode was held between two bifunctional oxygen electrodes. The capacity of the 25 cm^2 zinc electrode ($\varnothing$ = 56.4 mm, thickness ca. 1.4 mm) was 2.5 Ah. The electrode had two active sides and thus an effective surface area of 50 cm^2. The electrodes were prepared from a slurry containing 84 wt% ZnO, 4 wt% PTFE, 2 wt% PbO and 10 wt% Cellulose. The preparation procedure is discussed in detail in (7). Cellulose as a internal wick provides open electrolyte channels to keep the electrode wet and porous over long operation periods (8).

Dendritic shorting of the cell was prevented by wrapping the Zn electrode in three layers of Celgard 3401 separator (0,02 µm effective pore size, Hoechst Celanese Corp., Charlotte, North Carolina). Each separator layer was heat-sealed as close as possible to the edges of the electrode. Before the pasted zinc electrodes were mounted into the cell, they were degassed under vacuum and then soaked with 15 % KOH + 1.5 M KF + $ZnO_{sat.}$.

The formation of zinc on the electrode was realized by charging the electrode with 1.7 Ah (2/3 of the total capacity) at 1.25 mA/cm^2 and then discharging it using the same current density until the cell voltage dropped below 0.9 V. About three formation cycles were performed before the electrode was used for the cycle test. After the last formation cycle about 95 % of the charging current could be recovered in the discharge cycle. Under regular test conditions one third of the initial capacity was cycled at C/6 (6 hours charge rate) charge and C/3 discharge rates. The computer-controlled system (Software: Work Bench; Hardware: PC-AT 386 hooked up to Amel Model 555 B galvanostat) was programmed to allow constant current discharge until the zinc electrode potential became more positive than 300 mV vs. $Zn_{ref.}$ or until the cycled capacity (0.85 Ah) was reached. Between each half cycle the cells were exposed to a 6-min open-circuit period. Cycling continued until the cell capacity had definitely dropped below 50 % of its original value.

Bipolar Cell

A schematic presentation of the bipolar Zn/oxygen test battery is given in Figure 1b. In the bipolar system all cells are internally connected in series with a current take-off at each end of the cell stack. The current flows through the whole area of each electrode. This compact cell design has a low internal resistance, and batteries with high voltages can be obtained if the number of bipolar cells in the stack is high enough. Shunt current flowing through a common electrolyte manifold was avoided by using a separate electrolyte inlet/outlet for each cell.

For the bipolar battery the zinc electrode substrate was a porous copper foam material (specific surface area 4100 m^{-1}, geometric surface area 100 cm^2, thickness 2 mm) purchased from Sorapec (France). During the first charging, the electrolyte (45 % KOH + 0.18 M Zn(II)) flowed through the electrode and the zincate was reduced at the porous electrode. A initial capacity of 8 Ah was developed at 80 mA/cm^2. After the first charge, the electrolyte was replaced by a 20 % KOH solution presaturated with ZnO. During discharging in this moderately alkaline electrolyte, oxidized Zn precipitates as ZnO on the foam. There was no electrolyte circulation during cycling. The cycled capacity was 5 Ah. Cycling was performed at 10 mA/cm^2 and 20 mA/cm^2 charging and discharging currents, respectively. The power of the bipolar stack containing one to seven cells was calculated from measured current-voltage curves.

Bifunctional Oxygen Electrode

The two-layer (hydrophobic gas-permeable protection and semi-hydrophobic catalyst layer, respectively) bifunctional oxygen electrodes were fabricated by applying the rolling technique as described in (5). The active layer consisted of $La_{0,6}Ca_{0,4}CoO_3$-catalyst homogeneously dispersed on graphitized Vulcan XC 72 (graphitized at 2700 °C, He). The assembly was then sintered under a pressure of 50 kg/cm^2 at 340 °C.

RESULTS

Capacity of the Zinc Electrodes

In order to measure the recoverable capacity of the pasted zinc electrode as well as that of the zinc-foam electrode, we performed some galvanostatic discharge experiments at different discharge currents. For all these experiments the 25 cm^2 monopolar cell and the 100 cm^2 bipolar cell were charged with 1.615 Ah and 8 Ah, respectively.

Figure 2 (a and b) presents these discharge experiments by plotting the cell voltages against the cell capacity (taking 1.615 Ah and 8 Ah as 100 %, respectively). Both systems are showing qualitatively the same behaviour. For lower discharge currents the product distribution is more uniform and almost 100 % of the zinc formed during charge is available for the discharge process. At higher current densities concentration polarization and the decrease of pore size promote nonuniformity (9). For the porous zinc-foam electrode the resistance in the electrode matrix is less important than for the pasted zinc electrode. At current densities up to 100 mA/cm^2, more than 50 % of the zinc capacity of the porous zinc-foam electrode can be oxidized during discharge.

These results demonstrate that the battery may be discharged completely. For a long cycle life it is advantageous, however, to use an initial loading of Zn which is higher than the cycle capacity. In the pasted zinc electrode the relatively high amount of zinc not involved in cycling (one third of the nominal capacity) improves the conductivity of the electrode considerably. On the other hand, if the electrode is charged until 100 % of the ZnO is transformed to Zn, the porosity of the electrode may decrease substantially upon cycling. For routine cycling a current of 0.3 A (6 mA/cm^2; active geometrical surface area = 50 cm^2) and a capacity of approx. 0.85 Ah were chosen for an electrode with 2.5 Ah capacity. Under these cycling conditions, and taking into account the electrode and electrolyte mass of our laboratory cell combined with an estimated mass for the cell case and hardware (approx. 15 % of the complete battery mass), a specific energy of approx. 70 Wh/kg was calculated. Approximately the same specific energy was estimated by applying Zn-foam and a cycle capacity of 5 Ah in the 100 cm^2 electrode.

Specific Peak Power

Monopolar Cell. The maximum power of the monopolar cell was measured at different states of charge. Figure 3a shows the cell characteristics measured with a pasted zinc electrode containing (a) 66 % (charged cell), (b) 49 %, (c) 37 % and (d) 28 % metallic zinc. The voltage of the cell was measured for each particular current over a time period of 30 s. The increasing amount of ZnO is leading to a decrease in conductivity in the zinc electrode, which is indicated by the change of the slope in the current-potential curves. After recharging of the cell (66 % Zn/33 % ZnO) the cell characteristics described could be reproduced. The calculated specific peak power (based on the same weight assumptions as used for the specific energy calculation) was 160 W/kg for the charged monopolar cell.

Bipolar Stack. A compact bipolar cell arrangement is interesting in order to diminish the ohmic resistance of the module and to obtain high specific power. On the other hand for the bipolar zinc-oxygen battery only one bifunctional oxygen diffusion electrode faces each zinc anode. Consequently the power output of the bipolar stack depends even more on the oxygen electrode properties. The power curves for the bipolar stack containing one to seven cells are plotted in Figure 3b. Due to the absence of shunt currents and the use of electrodes with very similar electrochemical behaviour, the experimentally measured voltage of the battery was very close to the expected value ($U_{bipolar\ stack} = N*U_{one\ cell}$). For the bipolar stack with seven cells a peak power of 90 W (130 mW/cm^2) could be demonstrated. Comparing the power densities [in mW/cm^2] of the monopolar and bipolar cell we can see that there is no peak power limitation due to oxygen electrode performance. The excellent conductivity provided by the porous Zn-foam electrode and its highly porous structure even in the charged state (ca. 80 % porosity) are responsible for the high peak power of the bipolar battery. While the power of the monopolar cell is reproducible at high cycle numbers, the bipolar battery shows a remarkable decay in performance within 10 cycles.

Cycle Life Performance

Pasted Zinc Electrode. In Figure 4 the cell capacity is plotted vs. the cycle number for a cell with an electrolyte containing 15 % KOH and 1.5 M KF + $ZnO_{sat.}$. The capacity of the cell dropped below 100 % after 100 cycles and remained at 90 % for the next 30 cycles. Capacity decay continued during the next 200 cycles. After 350 cycles the capacity dropped to 50 % of the design capacity (0.85 Ah). At this point a formation cycle at low rate (25 h, 1.26 mA/cm^2) was performed. After this formation cycle the cell delivered again an average capacity >60 %.

For the first 100 cycles the average charging and discharging voltages of this battery were 2.0 V and 1.2 V, respectively. About 90 % of the potential losses in this cell are related to the overpotential at the bifunctional oxygen electrode. Its deviation from the standard potential was approximately 250 mV for the oxygen evolution and 450 mV for the reduction process. The polarization at the Zn electrode was close to 50 mV.

At the beginning of the cycle test the calculated energy efficiency was about 57 %. However, the average energy efficiency calculated over 400 cycles was ca. 45 %. A small overcharge of about 3 % was necessary to compensate for zinc corrosion. Dendritic shorting was not observed during the long-term test. It seems that any zinc dendrites produced during charging are oxidized in the O_2-rich environment.

X-ray photographs of the pasted zinc electrode still showed regions of dense zinc in the completely discharged electrode after the long time experiment for (discharging at low rate as applied during the formation cycles). Passivation of the metallic zinc could occur by physical blocking of the active zinc surface by a precipitated ZnO film. Porosity measurements of the pasted zinc electrode before and after the cycle test supported the speculation of zinc passivation induced by densification of the porous electrode at high cycle numbers. During cycling the porosity of the pasted Zn electrode markedly decreased from 50 % to 20 %.

Zinc-Foam Electrode. To achieve a homogeneous zinc deposition on the reticulated Cu foam, the first charging process was performed at limiting current density and high flow rate of the 45 % KOH electrolyte. For the cycling test stationary 20 % KOH solution was applied. In Figure 5 the cell voltage of a single cell (100 cm^2 end-plates of the bipolar battery) are plotted versus the cycle number. The slightly lower energy efficiency of this system can be explained by the fact that the applied current densities for the charging and discharging process were ca. three times as high as for the pasted zinc electrode. Under these conditions the cell performance decreased within the first 17 cycles. The higher polarization of the cell during cycling is due to an increase in polarization of the zinc-foam electrode. Visual observations showed redistribution and dissolution of the metallic zinc in the foam electrode. After perfoming a formation cycle (same conditions as for the first charging process) the cell could be cycled for another 17 cycles.

CONCLUSIONS

The combination of $La_{0.6}Ca_{0.4}CoO_3$-catalyzed bifunctional oxygen electrodes with pasted zinc electrodes and moderately alkaline electrolyte led to a long-life rechargeable battery. 450 charge/discharge cycles and a peak power of ca. 160 W/kg were determined for 0.85 Ah (cycle capacity) zinc-oxygen cells. In order to obtain long cycle life and a specific energy of 70 - 90 Wh/kg, only one third of the nominal capacity of 2.5 Ah should be cycled at C/6 charge and C/3 discharge rates. At useful discharge rates for electric vehicle applications (C/1 - C/2) the specific energy of the battery is still ca. 60 Wh/kg if only one third of the nominal capacity is cycled.

The bipolar battery research was focused on the demonstration of a high specific power. With a bipolar stack of seven cells a maximum power of 90 W could be measured. The porous zinc-foam electrode gave higher power densities than the pasted zinc electrode. On the other hand, redistribution and dissolution of the zinc induced by natural convection, resulted in a limited cycle life. At periodic service intervals the formation of new, well distributed zinc on the foam has to be performed while providing fresh flowing alkaline electrolyte.

ACKNOWLEDGMENTS

The authors acknowledge the financial support of this work provided by the Swiss Federal Office of Energy. We also thank Mr. S. Boss and Mrs. G. Masanz for excellent technical support.

REFERENCES

1. R. Putt, Zinc-air Design Concept for the DOE-EHP IDSEP Van, Matsi Inc., July 31, 1990.

2. T. C. Adler, F.R. McLarnon and E. J. Cairns, J. Electrochem. Soc., **140**, 289 (1993).

3. Y. Shimizu, K. Uemura, H. Matsuda, N. Miura and N. Yamazoe, J. Electrochem. Soc., **137**, 3430 (1990).

4. P.N. Ross and M. Sattler, J. Electrochem.Soc., **135**, 1464 (1988).

5. S. Müller, K. Striebel, O. Haas, Electrochim. Acta, **39**, 1661 (1994).

6. C. Schlatter, C. Comninellis, S. Müller, O. Haas, Abstract, 45th ISE Meeting, Porto, 28 Aug - 2 Sep (1994).

7. R. Jain, M.S. Thesis, Lawrence Berkeley Laboratory, Report No. 25332, 1989.

8. S. Müller, F. Holzer and O. Haas, Abstract, 46th ISE Meeting, Xiamen, 27 Aug - 1 Sep (1995).

9. W. G. Sunu and D. N. Bennion, J. Electrochem. Soc., **127**, 2007 (1980).

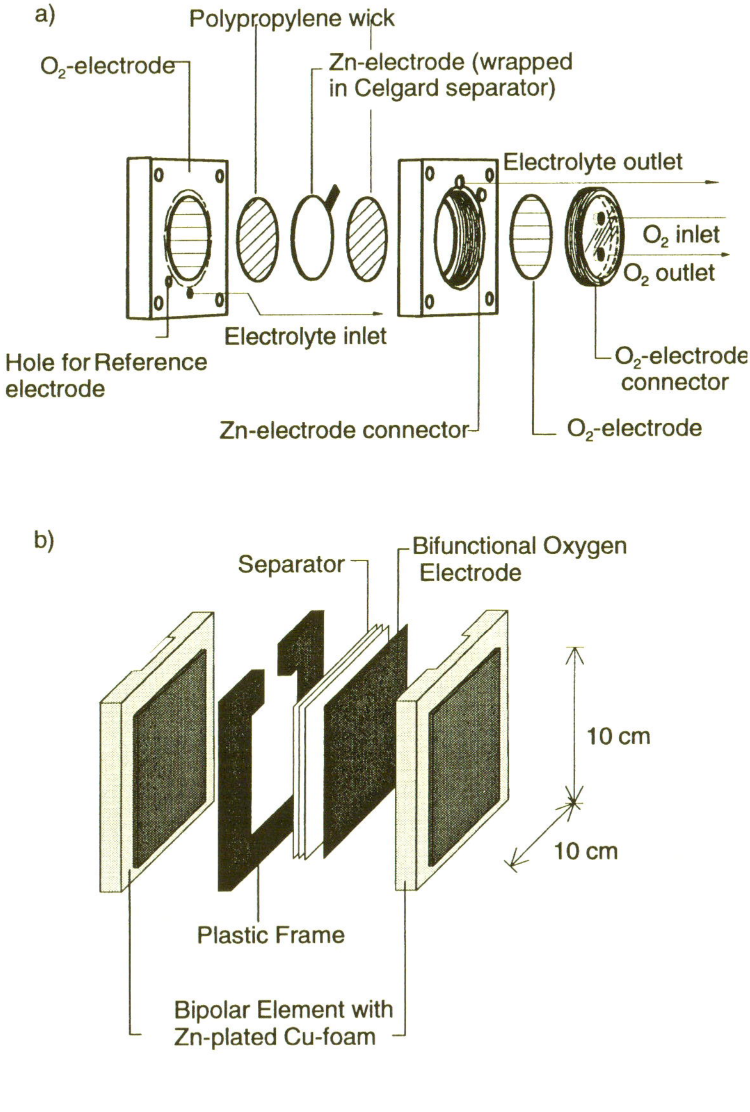

Figure 1 Exploded view of a) monopolar cell and b) bipolar stack components.

a)

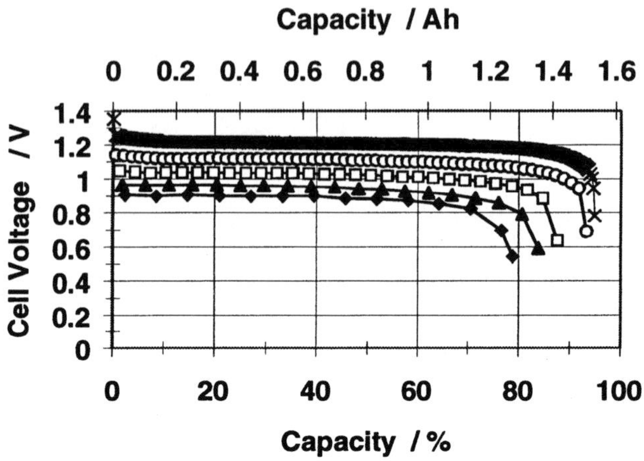

b)

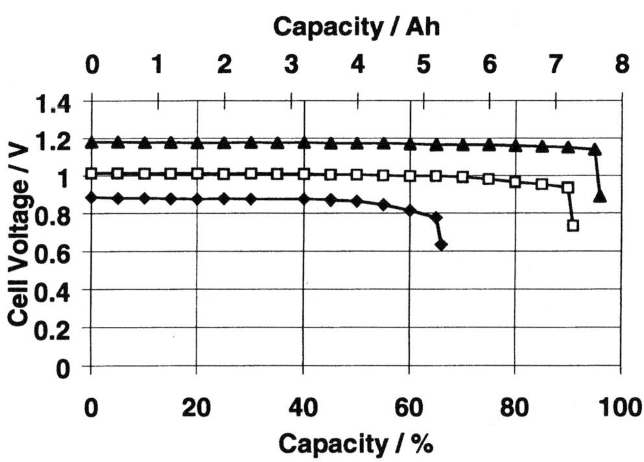

Figure 2 Zinc/oxygen cell voltages and available capacity for different rates of discharge measured for (a) a single 25 cm^2 cell with pasted zinc electrodes at (x) 0.064 A, (O) 0.3 A, (□) 0.6 A, (▲) 0.9 A, (◆) 1.2 A and for (b) at the 100 cm^2 end-plates of the bipolar stack at (▲) 1 A, (□) 4 A, (◆) 10 A.

a)

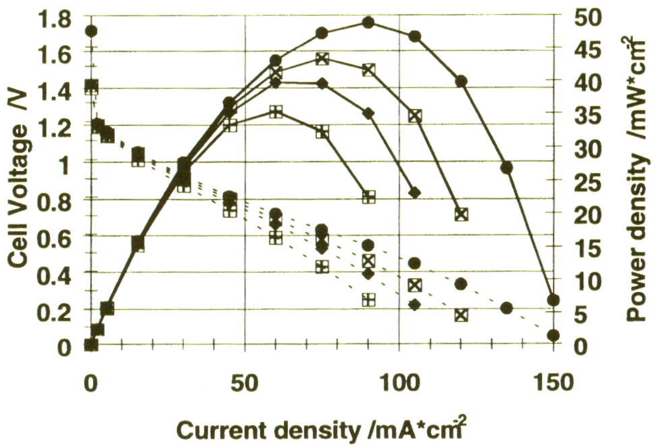

b)

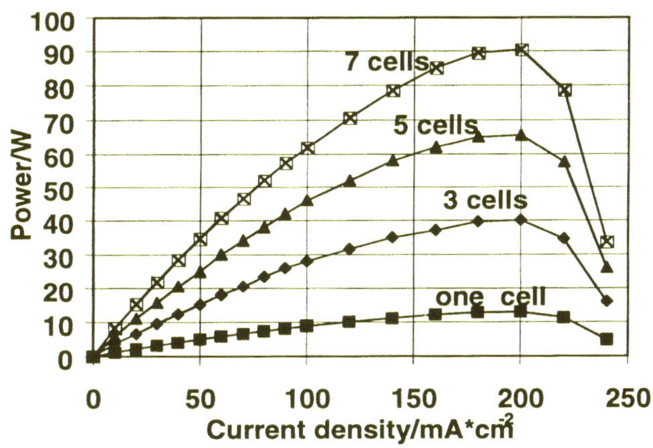

Figure 3 (a) Voltage, current density, and power density delivered by Zn/O_2 cells at different charge states: (•) 66 %, (x) 49 %, (◆) 37 %, (+) 28 % metallic zinc.

(b) Power curves measured for the bipolar stack containing one to seven cells (six bipolar plates and the end-plates). Electrolyte: 20 % KOH + $ZnO_{sat.}$.

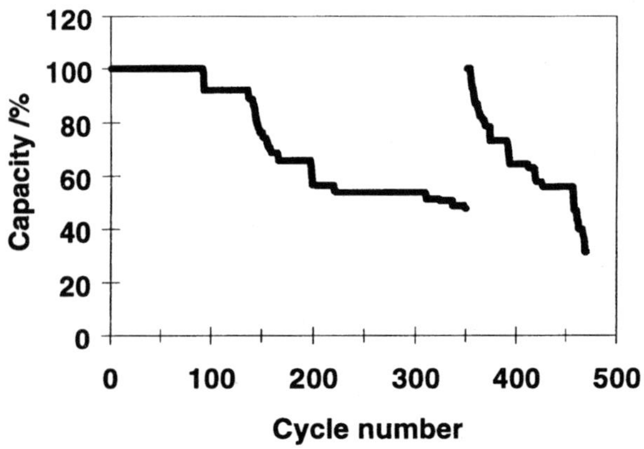

Figure 4 Cell capacity of the monopolar zinc/oxygen cell as a function of the cycle number. The average charging and discharging voltages are 2.0 V and 1.1 V, respectively. Electrolyte: 15 % KOH + 1.5 M KF + $ZnO_{sat.}$.

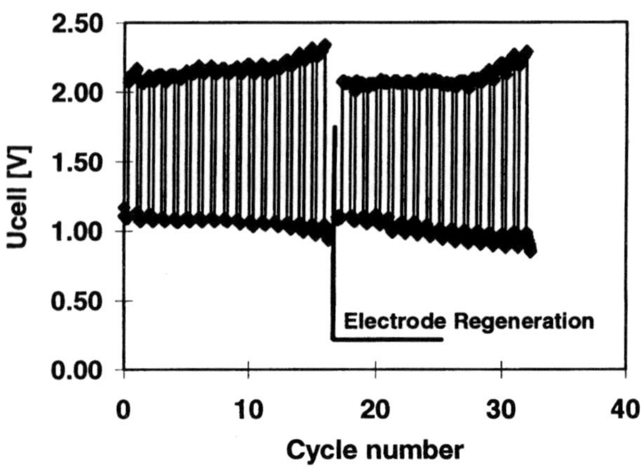

Figure 5 Votages of a single cell of the bipolar battery as a function of the cycle number. Charging and discharging currents are 10 and 20 mA/cm^2, respectively. Electrolyte: 20 % KOH + $ZnO_{sat.}$.

POWER AND PERFORMANCE CHARACTERISTICS OF RECHARGEABLE ZINC-AIR BATTERIES FOR LONG RUN TIME PORTABLE POWER APPLICATIONS

M. N. Golovin, L. A. Tinker, E. S. Buzzelli
AER Energy Resources, Inc.
1500 Wilson Way, Suite 250
Smyrna, GA 30082

ABSTRACT

AER Energy Resources, Inc. has developed a long run time zinc-air rechargeable battery system for portable power applications. The battery system uses cells employing two bifunctional air electrodes, a single zinc electrode, and KOH based electrolyte. The cells provide greater than 170 Wh/kg and 200 Wh/l at C/8 discharge rate. An air management housing completes the battery system. The air manager is designed to provide a uniform air supply to the cells and minimize effects of environmental exposure. Performance data with respect to power, run time per cycle, recharge efficiency, and cyclic stability of cells utilized in the battery system will be discussed.

INTRODUCTION

Primary zinc-air battery technology has been in commercial use for many years (1). The most common application today is in small button cells for hearing aids. The zinc-air system provides high energy density, a flat discharge voltage, long shelf life (sealed), and relatively low cost components. Portable power applications such as laptop computers require rechargeable batteries to provide service life and convenience in use. A limitation in convenience, however, is that run time between charges is on the order of two to four hours (2), significantly shorter than a typical work day. In order to achieve longer run times between charges, batteries must provide higher energy storage capability without large sacrifices in weight and volume. Zinc-air technology can provide the high energy storage capability required for long run times and its other characteristics make it attractive for such applications.

Development of electrically rechargeable zinc-air technology has been limited by the need for an efficient bifunctional air electrode and the apparent limited rechargeability of the zinc electrode. AER Energy Resources, Inc., has spent the last five years developing an electrically rechargeable zinc-air cell technology that can meet the long run time needs of the portable computer industry. The battery systems developed employ zinc-air cells combined with an air manager system to provide long run time solutions for portable computer users. This paper discusses the characteristics of the cell technology utilized in these designs and presents typical performance data.

CELL DESIGN AND PERFORMANCE

Design

A cross section of the AER Energy zinc-air cell design is shown in Figure 1a. The cell is a dual air electrode design making use of both sides of the zinc electrode. The cell case is injection molded polypropylene with air openings sized for the discharge rate required at the system level. The cells operate at ambient pressure and the ends of the cells contain a low pressure vent system that prevents excessive internal pressure buildup. The air electrodes are carbon based electrodes with a Teflon® binder and contain a current collector, a hydrophobic layer, and an active hydrophillic layer. The hydrophillic layer contains catalysts that assist in the discharge and charge process. A single zinc electrode is used and the separator package consists of microporous polypropylene and nylon based materials. The zinc electrode is the capacity limiting electrode. A 38% by weight potassium hydroxide solution is used as the electrolyte. A complete Model 13220 cell design is shown in Figure 1b. This cell has 132 cm^2 of air electrode area and will deliver 26 Ah of discharge capacity to 0.9 volts with an average discharge voltage of 1.1 volts. The cell weighs 0.165 kg and has a volume of 135 cm^3. This translates to a specific energy of >170 Wh/kg and an energy density of >200 Wh/l.

Component Performance

The negative electrode of the zinc-air rechargeable system is the zinc electrode. The discharge and charge reactions for the this electrode are shown in equations [1] and [2].

$$Zn + 4\ OH^- \Leftrightarrow Zn(OH)_4^{=} + 2e^- \quad [1]$$

$$Zn(OH)_4^{=} \Leftrightarrow ZnO + 2\ OH^- + H_2O \quad [2]$$

On discharge zinc is converted to zincate ion, $Zn(OH)_4^{=}$, which then forms ZnO after reaching a saturation point that can be as high as 300% of the equilibrium saturation level (3). On charge the dissolution of ZnO precedes the reduction of the zincate ion to reform the zinc electrode structure. Because the reactions involve soluble intermediates and precipitation of non-conductive ZnO, corrosion and passivation of the zinc electrode are important considerations. The main reason for concern about corrosion and/or passivation is that such processes can limit the available anode active material. Figure 2 shows the corrosion/passivation chemistry for a typical zinc electrode. The data were collected on a 7 cm^2 piece of zinc anode material using a Zn metal (99.98%, Alfa) quasi-reference electrode (QRE) and a Ni mesh counter electrode. Figure 3 shows the Tafel region of the data in Figure 2 along with the fit to the Tafel equation. The data show that the E_{CORR} for the Zn anode is -10 mV *vs.* Zn with a corrosion current of 41 $\mu A\ cm^{-2}$. The Tafel data and slopes are fairly consistent with those reported in the literature (3). The data also indicate that these issues are of minor importance under conditions seen in the cells described here.

The positive electrode is a bifunctional air electrode that reduces oxygen from the air to produce OH^- ions on discharge and oxidizes OH^- ions to produce oxygen on recharge. A simplified reaction ignoring intermediate species is shown in equation [3].

$$O_2 + 2\ H_2O + 4e^- \Leftrightarrow 4\ OH^- \quad [3]$$

A cyclic voltammogram of the air electrode is presented in Figure 4. These data were obtained in a 1 cm^2 cell using a Zn reference and counter electrodes with 38% KOH electrolyte. The voltage excursions at the higher and lower voltage extremes are the oxygen evolution and reduction reactions, respectively. The anodic and cathodic waves seen between are attributable to the peroxide removal catalyst.

The data in Figure 4 suggest that the air electrode should be capable of sustaining 120 mA cm^{-2} at 1.1 V. In practice, however, the air electrode will support about 15 to 20 ma cm^{-2} at 1.1 V for extended discharges. This is illustrated in Figure 5 where a polarization curve obtained on a production cell at room temperature with a Hg/HgO reference electrode inserted is shown. The data indicate that at discharge rates between 1.0 and 4.0 A (7.5 to 30.3 mA cm^{-2}), the air electrode voltage polarization controls the capability of the cell. Stability of the air electrode has been demonstrated by removing the air electrode of a cell cycled for over 350 hours (32 cycles), and reusing it in a cell with all other components being fresh. This cell cycled with only a slight loss of performance most likely attributable to handling during rebuilding.

Prototype Cell Performance

Prototype cells of the design described above have been extensively tested. The discharge profile for a typical cell is displayed in Figure 6. The important point to note is that the voltage profile is a fairly flat one. In constant power applications, such as lap-top computers, this flat voltage profile at constant power is an advantage. The charge profile for the same cell is shown in Figure 7. The cell was charged to an amp-hour endpoint such that all of the discharged capacity was replaced. The charge voltage is about 2.0 V and is somewhat sloped. The charge can be voltage limited or a taper charge regime may be used. The present charging method is used primarily to study the coulombic efficiency while more effective methods of charging are used in batteries.

The data in Figure 8 show the capacity maintenance and overall run time for a single cell discharged to 100% depth of discharge (DOD) along with the cumulative discharge hours for the cell. The data clearly show that though the cell cycled only 32 discharge/charge cycles, the cell delivered a cumulative discharge life of over 400 hours. The cell also cycles with high coulombic efficiency for a rechargeable zinc system. The data in Figure 9 show that the cell can operate with >95% efficiency for over 80% of its operational life with the anode (zinc electrode) being the primary source of inefficiency. The cell chemistry is also very stable. The average discharge voltage of the cell (Figure 10) did not decline until the last few hours of discharges. The voltage increase in the first 50-100 hours of operation are the breaking in of the cell chemistry. The cell then stabilized after about 200 hours of operation and the cell shows a rapid decline after 450 hours of operation. The rate capability of the cell is shown in Figure 11. The data were collected by discharging a standard cell at different rates with a power supply and allowing the cell voltage to stabilize. The data show that the cell can maintain voltages of around 1.1 V at 16 mA cm^{-2} (2 A at 132 cm^2). The cell is clearly a highly efficient, rechargeable system capable of delivering the desired run time at a reliable voltage.

Production Cell Performance

Rate capability and temperature dependence data are shown for typical production cells in Figures 12 and 13, respectively. The data in Figure 12 show that the main difference in the voltage profile for the different loads is the effect of cell impedance. The voltage profiles are all roughly parallel, showing a slight decrease in average cell voltage as the rate of discharge was increased. A similar argument can be made for the discharge of a production cell at 4, 21, and 40 °C (Figure 13). The profiles at the different temperatures are all fairly parallel, with a decrease in average voltage as the temperature is decreased. These data show that the cell chemistry for the rechargeable zinc-air cell is stable giving a constant voltage discharge which is only affected by the cell impedance.

At the battery system level cells are limited to a depth of discharge of about 80% of the zinc capacity or 20 Ah and are operated by design at less than the rate limiting condition. This will provide more than 8 hours of operation in typical applications. Figure 14 shows the typical cycle life performance of cells discharged to a 20 Ah capacity limit. The data show that the cell can deliver over 300 hours, or 600 Ah in this case, of continuous operating life. It is important to remember that the end-user does not acknowledge cycle life, but rather does know when the battery of an application such as a portable computer is not delivering the run time that is desired. Thus, total run time is a more important parameter to look at than is the number of discharge/charge cycles.

SUMMARY

The performance features of rechargeable zinc-air cells designed and built at AER Energy Resources, Inc. have been described. The data demonstrates that the zinc-air system has high coulombic cycle efficiency and is stable over a service life sufficient to support the intended application. In addition this technology can provide long run time between charges because of its high energy storage capability. Work is continuing to improve the energy density, power capability, and total operational life of this technology. Production cells like those described are being built into accessory batteries for portable computer applications.

REFERENCES

1. S. F. Bender, J. W. Cretzmeyer, and T. F. Reise, in Handbook of Batteries 2nd Edition, D. Linden, Editor, p. 13.1, McGraw-Hill, New York (1994).

2. B. Howard, in PC Magazine, Vol. 14, No. 14, p. 102, Ziff-Davis, New York (1995).

3. M. J. Isaacson, F. R. McLarnon, E. J. Cairns; J. Electrochem. Soc., Vol. 137, No. 7, p. 2014 (1990).

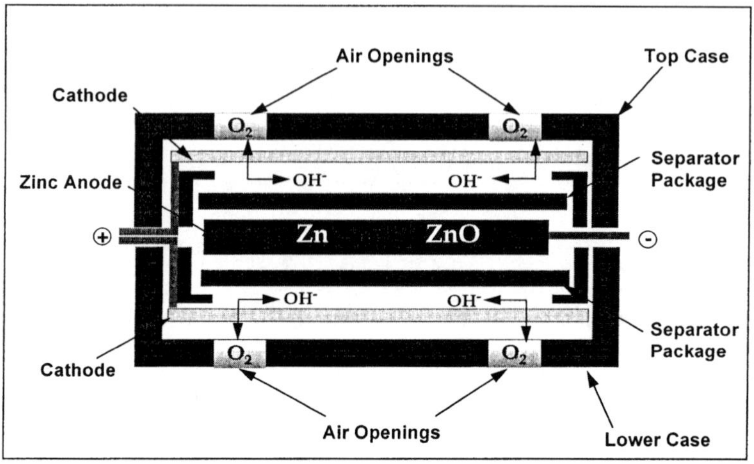

Figure 1a: Cross Section of Dual Air Electrode Zinc-Air Cell

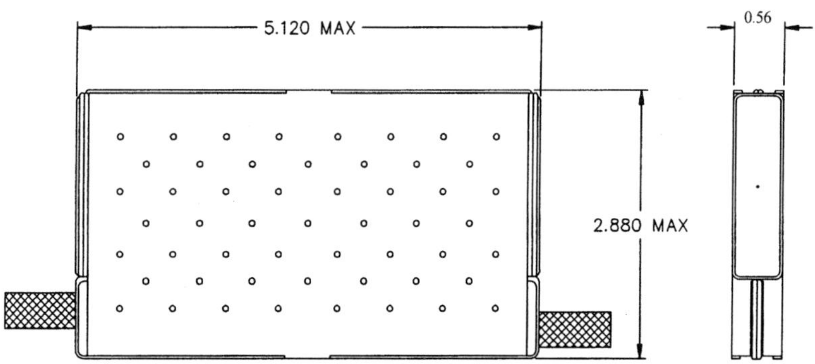

Figure 1b: Model 13220 Dual Air Electrode Zinc-air Cell Design

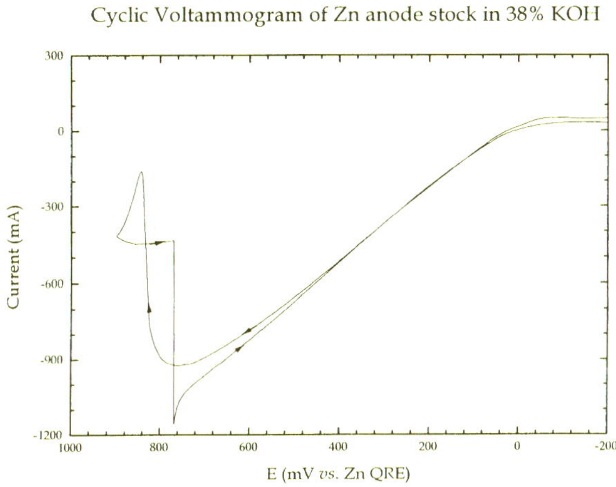

Figure 2: Cyclic voltammogram of zinc anode material in 38% KOH obtained at 10mV s^{-1}. A Ni mesh counter and a zinc reference electrode were used in a beaker type cell.

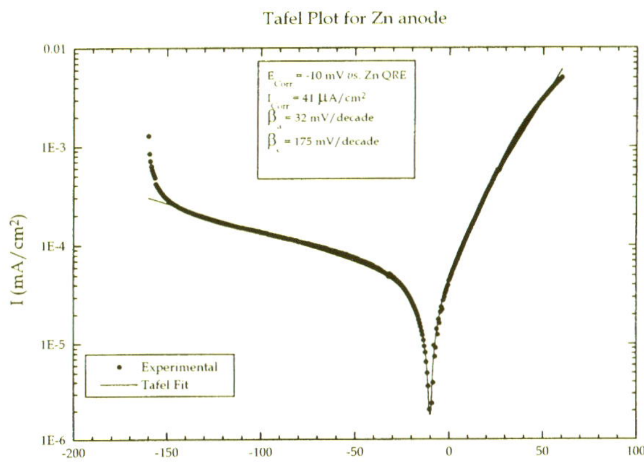

Figure 3: Tafel Plot for the corrosion of zinc anode material in 38% KOH

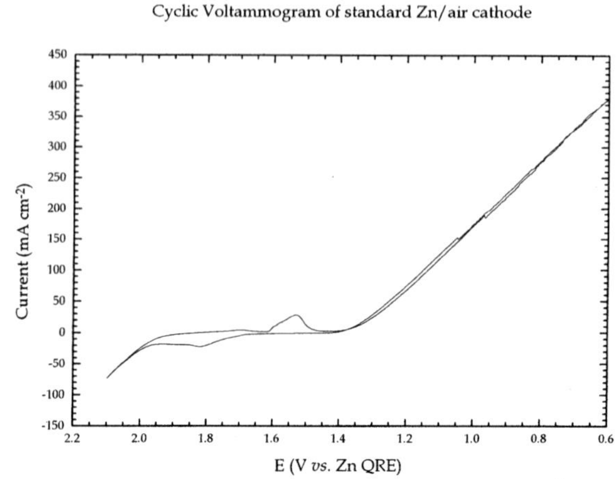

Figure 4:Cyclic Voltammogram of a standard cathode at 4 mV s^{-1} in 38% KOH. A Zn reference and counter electrode were used

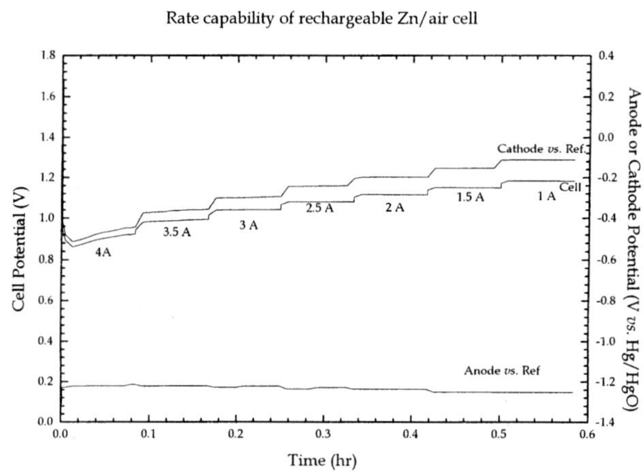

Figure 5: Rate capability for a rechargeable Zinc-air cell and the polarization of the anode and cathode under load.

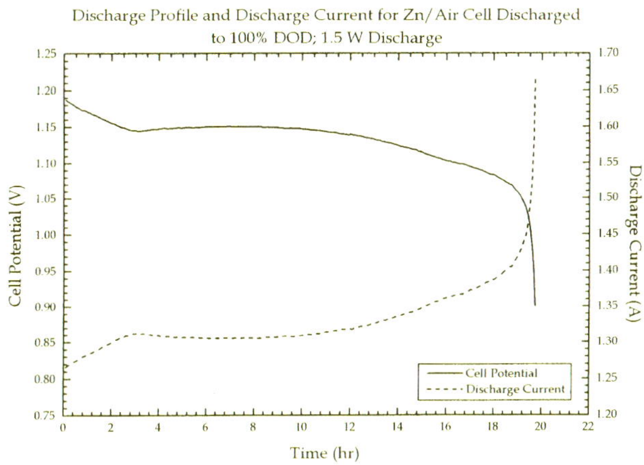

Figure 6: Discharge voltage profile for a rechargeable Zinc-air cell discharged to 100% DOD at 1.5 W

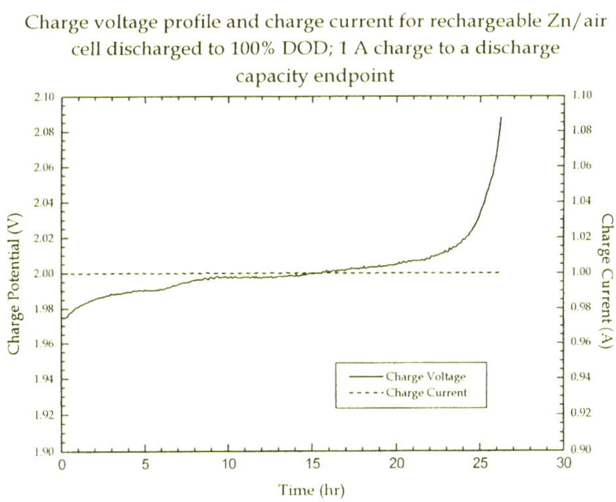

Figure 7: Charge voltage profile for a rechargeable Zinc-air cell charged to 100% of the previous discharge capacity at 1.0 A

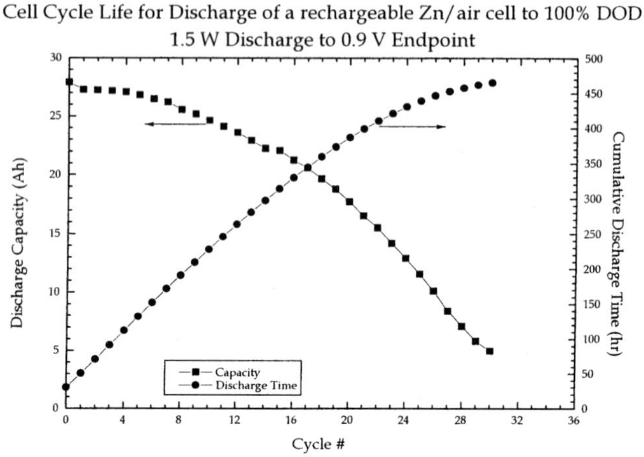

Figure 8: Discharge capacity and cumulative discharge time for standard Zinc-air rechargeable cell at 100% DOD

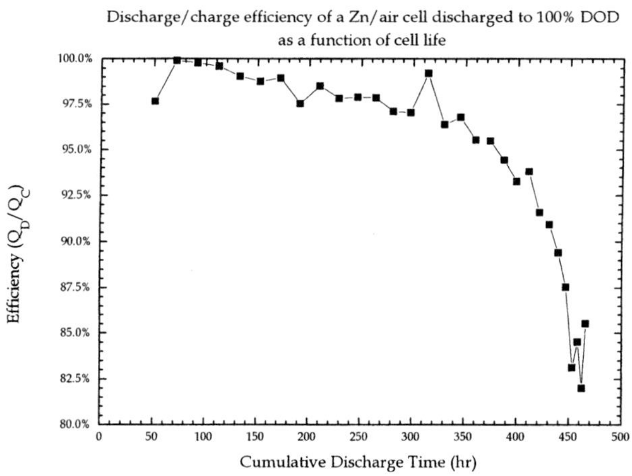

Figure 9: Coulombic efficiency as a function of cell life for a rechargeable Zinc-air cell discharged to 100% DOD at 1.5 W

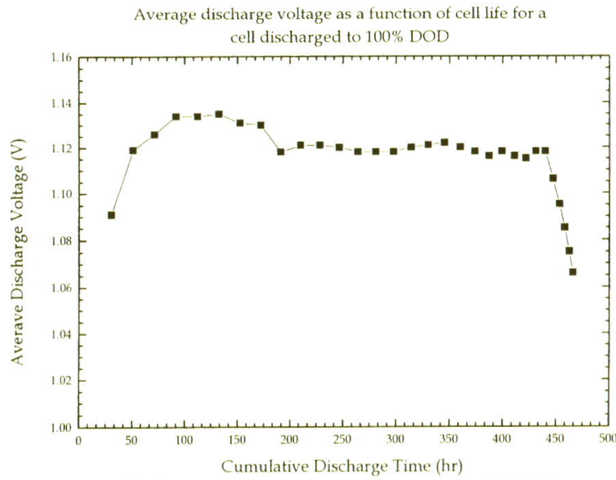

Figure 10: Average discharge voltage as a function of cell life for a rechargeable Zinc-air cell discharged to 100% DOD at 1.5 W

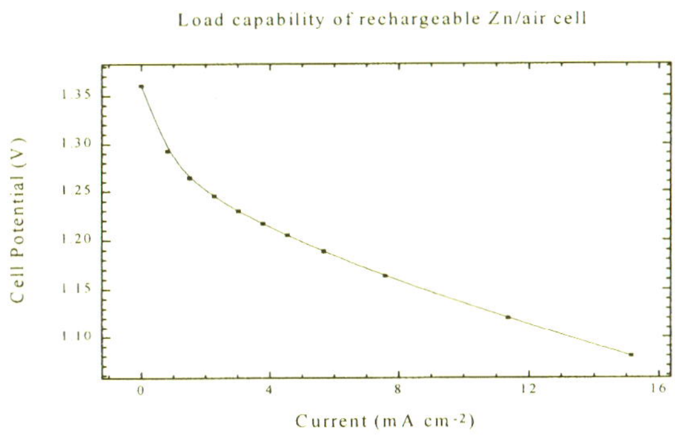

Figure 11: Polarization Data for a rechargeable Zinc-air cell

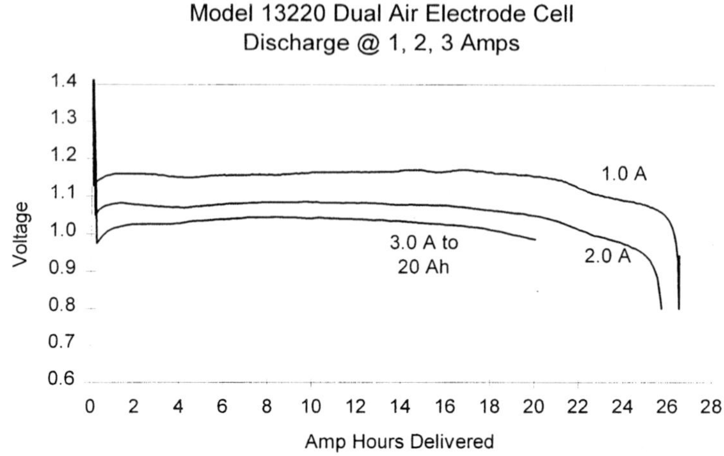

Figure 12: Cell Capacity versus Discharge Rate

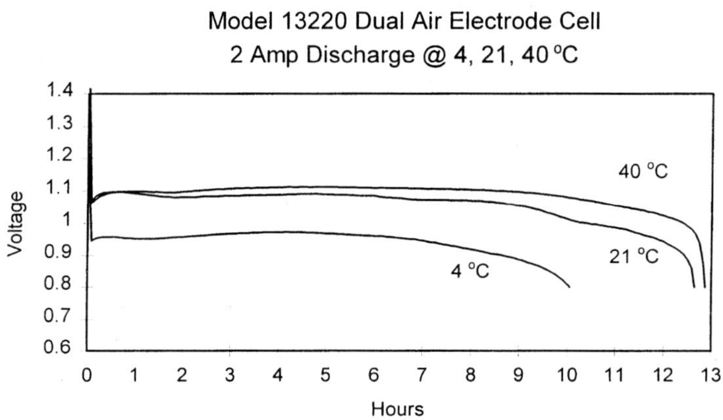

Figure 13: Temperature Effect on Discharge Voltage and Capacity Delivery

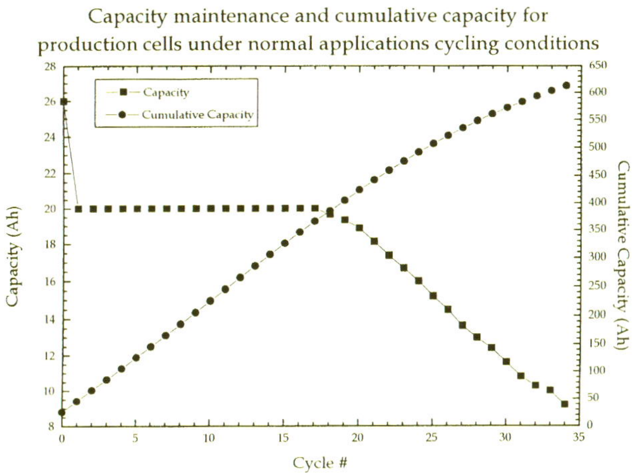

Figure 14: Cycle life for a production cell cycled at 2 A discharge to 20 Ah or 0.8 V and 2 A charge to 1.99 V with taper

DIFFUSION LIMITATIONS TO CYCLE LIFE OF NICKEL-ZINC BATTERIES AND EFFECTS OF VIBRATION AND ROTATION

Alvin J. Salkind (1,2)
Edward Kantner (1,2) and
Charles Grun (1,2)

1. Rutgers University, College of Engineering, P.O. Box 909, Piscataway, 08855-0909, USA

2. UMDNJ- Robert Wood Johnson Medical School, Piscataway, NJ 08854, USA

ABSTRACT

The relatively short cycle life of zinc electrode battery systems is mainly attributed to the lack of restoration of the electrode to the same morphology when recharging, Vibrating and or rotating the electrode during the recharging process has been shown to greatly improve cycle life and performance. The same beneficial results have been reported to be obtained by pulsing the electrolyte or other methods of minimizing the diffusion layer surrounding the electrode.

INTRODUCTION

The Nickel-Zinc battery system has been of scientific interest for almost a century and of commercial interest as a secondary battery for more than 60 years [1-4]. The advantages of a high level voltage of 1.60-1.65V, high power density, reasonable cost, and compact volume attracted research and development in many parts of the world. The most severe drawbacks to the system have been limitations to cycle life and difficulties in the design and fabrication of sealed batteries. A graphic illustration of the gravimetric and volumetric utilization of a typical design are shown in Figure 1, 1A and 1B.

The first large scale attempt at commercializing the system was the activities of Prof. James Drumm in the late 1920's. He had been given an initial grant of L 5000 by the Irish Free State parliament to work on silver additive modifications to the cathodes with the purpose of achieving electric rail travel. However, the positive electrodes actually used for the batteries that went into service were of the standard Jungner type, i.e. 23 mm wide pockets and filled with thick briquettes supplied by the company Alkaline Batteries Ltd in Redditch England [5]. The electrolyte did not contain lithia which was commonly used at the time by Edison and other fabricators of Nickel electrode systems as documented in Fig. 2. Drumm was reported to be a talented research chemist who had previously achieved practical success with methods of preserving green peas [6] and of devising an effective shaving soap. There were excessive and glowing claims made for the system by the sponsors, which is reported to have caused indignation and annoyance to Drumm. The objective of the Drumm patent [7] was to provide a lighter, more compact, and more conductive positive compared to the bulky pocket type nickel electrode in use at the time. The specific claims of the inventor are given in Table 1. After more than a decade of Irish government subsidies, the Drumm Battery Company operations were suspended in 1941 and apparently were never resumed.

Historically, the limits to cycle life of most zinc electrode systems has been changes in morphology and shape of the zinc electrode arising because of the high solubility of zinc discharge products in alkaline potassium hydroxide electrolytes. The zinc ions are free to migrate from their original source location in the matrix of the electrode. Consequently, when the zinc electrode is regenerated during the charging period, the diffusion parameters of zinc ions in electrolyte solutions result in progressive non-uniform deposition. This effect is variable with the nature of the electrolyte and other parameters which control diffusion. Changes to the electrolyte composition have been explored as a method of reducing the solubility of zinc, with limited success. The reaction at the zinc negative electrode during self discharge or use is:

$$Zn + 4OH^- \dashrightarrow Zn(OH)_4^{2-} + 2e$$
$$Zn(OH)_4^{2-} \dashrightarrow ZnO + 2OH^- + H_2O + 2e$$

These reactions imply the dissolution of the zinc electrode, with the crystallization of zinc oxide from the electrolyte. Once this mechanism is understood, the reaction at the anode can be simplified to:

$$Zn + 2OH^{-} \text{-->} ZnO + H_2O + 2e$$

This simple equation does not provide an understanding of the complications of a soluble intermediate. The problems of cycle-life and shape change were reviewed by McBreen [8].

Other problems with some designs has been the tendency of some types of nickel electrodes to be poisoned by the zincate electrolyte. M.C. Tsigankov [9] and V.N. Flerov [10] found that sintered nickel type electrodes were considerably less sensitive to zincate poisoning. Flerov's conclusions are shown in Table 2. His discussion also includes thoughtful comments on the nature of the interaction between a powder zinc electrode, contained in a cellophane envelope, and the zinc concentration at which the precipitation of the solid phase from the electrolyte begins.

In prismatic Nickel-Zinc cells, the main effects commonly observed on cycling are dendrite formation and capacity losses associated with the progressive imbalance in the distribution of active material are accelerated when the charge is extensive. The relatively poor charge efficiency of nickel electrodes at low currents shown in Fig. 3 compared to the high charge efficiency of zinc, creates a need for overcharging. Therefore, charge and discharge profiles can have a very substantial effect on cycle life.

BACKGROUND

The effects of diffusion on the anodic dissolution of zinc was observed and discussed in many reports. The work of E. Frankowiak [11] established the exact relation of anodic overpotential to reciprocal current in Levich Plots for both pure zinc pellets and zinc pellets with 5% acetylene black. This methodology and presentation serve as an extremely useful background to the understanding of the techniques for improving the cycle life of the zinc electrode. The data reported is replotted as figure 4. The author concludes that acetylene black [AB] acts as a sponge for the reaction products and for the electrolyte strongly limiting their diffusion to the bulk. The assumption was made that the mechanism has been effected and that reactions take place in the solid state.

In earlier work reported from the Lucas Research Centre in England, data or the use of graphite in the formation of porous zinc electrodes is detailed. This indicated that up to 32% by weight graphite was helpful as shown in the attached figure 5. [Ref. 25]

A review of the problems of zinc as a secondary battery electrode was prepared by T.P. Dirkse [12] in 1970. Earlier reports by Stachurski, Dalin, and Solomon [13,14] and others confirmed that dendrites form at potentials where diffusion limiting currents begin, whereas mossy or smooth deposits where activation energy controls that plating. This recognition lead to many types of cell designs intended to modify or overcome the diffusional problems of the Nickel-Zinc system. It has been shown that dendrite formation becomes significant at current densities above 20ma/cm2 and at overpotentials greater than 75mv.

NICKEL-ZINC DESIGN VARIATIONS

A unique design of a nickel-zinc battery cell which appeared to overcome the diffusional limitations and cycle life limitations was proposed by Pentti Tamminen [15, 16]. He performed studies showing the change in zinc electrode replating morphology with rotation as shown in figure 6. His reactor is shown in Figure 7. The battery is round, the plates horizontal and the system is bipolar. A slight rotation of the electrodes was given during charge and the electrolyte was capable of flowing from the central top feed point, dripping over the plate edges, and collecting in a reservoir at the bottom of the batter. Although the design sounds complicated, the reactor was very compact. A number of reactors of various sizes were fabricated and some cycled for more than 1000 cycles. In full size multicell batteries, specific energies of 70-80 wh/kg were achieved. However, the perceived drawbacks of flowing electrolyte and rotating equipment has stopped commercialization of this design so far. (It may be useful for other electrochemical functions such as waste water treatment.) The effect of nickel electrode thickness and efficiency on this design is shown in figure 8. This effect is almost the same for all Nickel-Zinc designs.

Another approach to changing the apparent electrolyte diffusion film thickness on the zinc was the design proposed by Otto Von Krusenstierna [17] called "Vibrocel". In this design conventional prismatic cases were used as shown on Figure 10. His concepts and preliminary data is shown in Figure 9. The nickel electrodes were of the sintered type and were connected by flexible straps to the positive terminals. The negative substrate was a woven wire mesh. The zinc was plated in place from a zinc rich electrolyte and was expected to dissolve during deep discharge. A number of experimental vehicles used this design in Sweden. Additional work was reported by E. Pearlman, D.T. Ferrell, and A.J. Salkind under Department of Energy Contracts [18, 19]. However, when closer plate spacings with reduced electrolyte quantities were attempted, in order to achieve high volumetric energy density, performance suffered. (Perhaps the same objectives of changing the microdiffusion in the electrolyte film surrounding the zinc active materials could be achieved by pumping the electrolyte).

Another variation of imparting vibration to the zinc electrode was patented by Kocherginsky, Kalachev, Naumenko, and Penkova in 1975 [22]. Their cell design is shown in Figure 11.

The other design variables mainly used to improve the cycle life of the nickel-zinc system involve:

- Separator Systems
- Modified Electrolytes
- Charging Systems

The various candidate separator materials for Nickel-Zinc cell were reviewed by Klein and McLarnon [20]. Their tabulation is shown in Table 3. There are variations and composite arrangements of these materials. The commonly available commercial materials are Celgard and PVA. However, most PVA's are slowly attacked by oxygen in the cell environment. Many current developments are directed at more inert materials.

The usual electrolyte for Nickel-Zinc cells is 20-35% KOH solutions. Other additives to the electrolyte include lithium hydroxide to preserve the conductivity of the nickel electrode and fluorides, carbonates, calcium hydroxide, and borate's to reduce zinc solubility and retard the rate and extent of zinc electrode shape change.

Charging considerations can have a dramatic effect on the life cycle performance of Nickel-Zinc cells. Since the nickel electrode needs more overcharge than the zinc, classic vented cells are normally constructed with an excess of zinc active materials. Charging schemes studied include:

- Modified Constant Potential
- Two Step Constant Current
- Superimposition of AC on DC

The latter is especially interesting as a method of maintaining the active materials surfaces in the not fully charged state. A program to study the effects of charge factors in Silver-Zinc cells for the LM Ascent Stage Battery was reported by Stoppler and Grun [21] in 1967. This work probably has significance in the operation of Ni-Zn cells. The asymmetric Charge Wave profile they used is shown in Figure 12. Their test cells were nominal 334 Ah units. Their best results were obtained with very low rate charges (100 hr rate) and the following parameters:

- I[DC] = 3.9 amperes
- I[AC] = 4.0 Amps, frequency f=30cps

Other studies specific to Nickel-Zinc cell charging were reported by Hennigan and Sizemore [24]. They used a two step voltage limit method. Other workers introduced an intermittent or on-off charging technique, which results in controlled diffusion layer thickness and allows decay of the concentration gradient which builds up during the "on" phase.

Generic studies on the pulse charging of zinc electrode systems were reviewed earlier in this proceedings by J. McBreen [23]. His statement that "pulsed charging has an enormous effect on zinc deposition in both free electrolyte and in pasted zinc electrodes" certainly has major implications in the development of long cycle life nickel-zinc cells.

RECENT ACTIVITY SUMMARY

There are two areas of research thrust for nickel-zinc cells now receiving much attention. Large cells for electric and Hybrid Vehicles and other propulsion applications, and small cells (mainly sealed) as a replacement for nickel-cadmium in limited cycle life electrical devices.

Recent and current activities involve immobilizing the zinc electrode, modifying the electrolyte to reduce zinc solubility, and/or limiting the diffusion limiting interface on the zinc electrode. This can be partially accomplished through control of the charging.

A small scale production of Ni-Zn cells has begun, mainly in Japan, for both electronic and EV applications.

FUTURE STUDIES AND WORK IN PROGRESS

- Effects of Surfactants
- New Separator Materials
- Sophisticated Electronic controls of Charging
- More charge efficient nickel electrodes
- New Electrolytes
- New hydrogen suppressing alloys and additives

REFERENCES

[1] S.U. Falk and A.J. Salkind, "Alkaline Storage Batteries", Wiley, NY (1969)

[2] F.P. Kober and A. Charkey, "Nickel-Zinc: A Practical High Energy Secondary Battery" in Power Sources 3-Proc. 7th Int Symp., Brighton, England (Sept 1970)

[3] V.S. Bagotsky and A.M. Skundin, "Chemical Power Sources", Academic Press, (1980)

[4a] A. Charkey, "Performance Characteristics of Nickel-Zinc Cells", 23rd Ann. Power Sources Conf. Atlantic City, (1969)

[4b] D. Reisner and M. Eisenberg, "A New High Energy Stabilized Nickel-Zinc Battery", SAE Tech Paper 890786 (1989)

[5] G.S. Atkinson, Alkaline Batteries Ltd. Redditch England, Private Communication to A.J. Salkind, (August 8, 1966)

[6] "The Drumm Battery", Discussion in 'The Electrical Review' August 22, 1930 p. 290-292
[7] J.J. Drumm, Brit. Pat. 365,125 (1930), U.S. Pat. 1,955,115 (1934)
[8] J. McBreen and E. Gannon, J. Power Sources, 15, 169 (1985)
[9] M.C. Tsignakov, Electrical Accumulators, Moscow, 1959 p. 21.
[10] V.N. Flerov, "The Behavior of Sintered Nickel Electrodes in Zincate Electrolyte", Zh. Prinkl. Khim, 32 (1959)
[11] E. Frankowiak, "Modification of Secondary Zinc Electrode", in Power Sources Symp. Bournemouth, Eng., (April, 1991)
[12] T.P. Dirkse, "Zinc as a Secondary Battery Electrode", Proceedings 24th Power Sources Symposium, (May 1990)
[13] Z. Stachurski, US Pat. 3,440,098 (April 22, 1969)
[14] Z. Stachurski, F. Solomon, and F. Dalin, Technical Reports from Yardney Electric Co.
[15] P. Tamminen, U.S. Pat. 4,521,497 and 4,684,585
[16] P. Tamminen and A.J. Salkind, "High Power Density, Long Life, Bipolar Nickel-Zinc Batteries", in Power Sources 12-Proc. Int. Power Sources Conf. 16, Brighton, Eng (1988)
[17] O. Von Krusenstierna, U.S. Pat 3,923,550 (1975)
[18] E. Pearlman, D.T. Ferrell, and A.J. Salkind, DOE Contract Reports on Nickel-Zinc Cells by Electric Storage Battery Co. (1976-1978)
[19] A.J. Salkind and E. Pearlman, "Alkaline Storage Batteries" in Kirk-Othmer Ency Chem. Tech., 3rd ed, Wiley Intersc. (1978)
[20] M. Klein and F. McLarnon, "Nickel-Zinc Batteries", Chapt. 29 in "Handbook of Batteries", Linden ed (1994)
[21] C. Stoppler and C. Grun, "Asymmetric Charging Cell Design", Research Report from Yardney Co. to NASA, (1967)
[22] M. D. Kocherginsky, L. Kalchev, A. Naumenko, and L. Penkova, U.S. Pat U.S. Pat 3,807,603, Sept. 23, 1975
[23] J. McBreen, "Zinc Electrode in Alkaline Solution", Proc. Electrochem Soc Symp. October 1995 and this volume.
[24] T. Hennigan and J. Sizemore, NASA reports
[25] Lucas Industries, U.K. Patent 2054252B (1983)

Table 1- Claims of Brit. Pat. 365,125, J. J. Drumm, (1930) [Ref 7]

An alkaline storage battery or accumulator, in which:

- Silver, or a silver compound, is employed as the positive active material.
- Finely divide silver is so used.
- Flaked graphite is mixed with the finely divided silver, or silver oxide.
- Nickelic Oxide is mixed with the silver, or silver oxide.
- Ceric Oxide is mixed with the silver, or silver oxide
- The positive plate is associated with a zinc negative.
- The positive plate is associated with a cadmium-iron negative
- The container is made of rustless, or nickel plated steel
- The container is made of "Monel" metal.

Table 2 - Abstract of Article by V.N. Flerov [ref. 10] "The Behavior of Sintered Nickel Electrodes in Zincate Electrolyte"

Conclusions:

- Sinter-plate nickel oxide electrodes are subject to the toxic effect of zincate to a much lesser degree than the pocket type electrodes.
- The working characteristics of the inter-plate nickel oxide electrode operating in the zincate electrolyte are considerably improved when the drying temperature is raised in the impregnation operation, when an addition of $Ba(OH)_2$ is introduced into the electrode and when a composite electrolyte is used in the accumulator.
- The toxic effect of zincate is increased with an increase in the amount of active material in the electrode. The preliminary forming of the electrode in the pure alkali begins to show a favorable influence in the latter case.
- Carbonation of the electrolyte is one of the basic causes of a reduction in the capacity of the sinter-plate electrode with the number of cycles.

***TABLE 3 - CANDIDATE SEPARATOR MATERIALS FOR NICKEL-ZINC CELLS**

Type	Thickness, cm Dry	 Wet	Resistivity, $m\Omega\text{-}cm^2$	Chemical Stability	Source
Cellophane	0.0025	0.0076	97-116	Poor	Dupont
Sausage casing	0.0076	0.00203	140	Poor	Union Carbide
PVA	0.0038	0.0076	225	Poor	Mono Sol
NASA-astropower	0.0305	0.0305	260	Good	Kimberly Clark
Electroporous	0.0050	0.0058	260	Good	Gould, Inc.
Celgard	0.0025	0.0025	260	Good	Celanese
RAI	0.0025	0.0025	260-390	Good	Radiation Applic Inc.
SAC	0.0050	0.0056	80	Good	Surface Activation Inc.
Microporous	0.00203	0.00203	775	Good	W.R. Grace
	0.00102	0.00102	225	Good	W.R. Grace
Nylon base	0.0050	0.0050	260	Good	ERC
PVA cross-linked	0.0025	0.0064	225	Fair	ERC

*From Klein & McLarnon [Ref. 20]

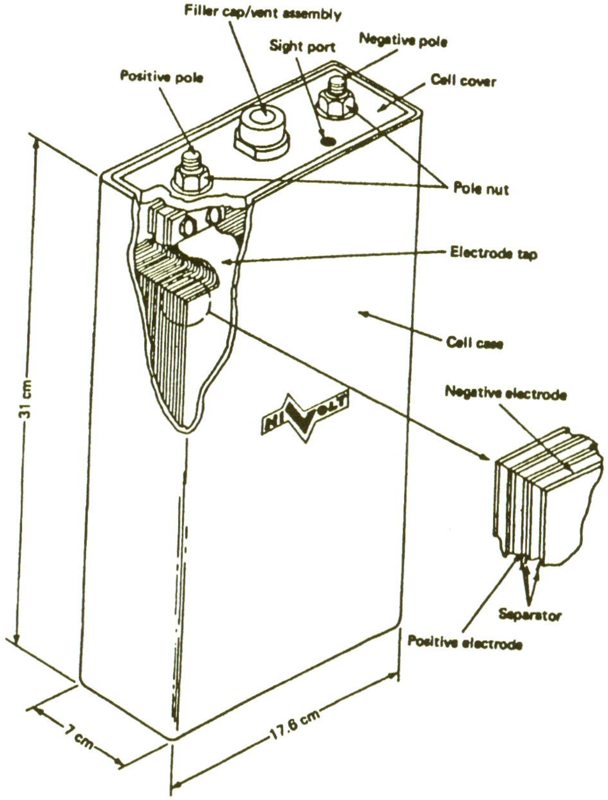

FIGURE 1 Nickel-zinc cell, 250 Ah. *(Courtesy of Energy Research Corp.)*

Fig. 1A Gravimetric Utilization of Nickel Zinc Cell (150 AH)

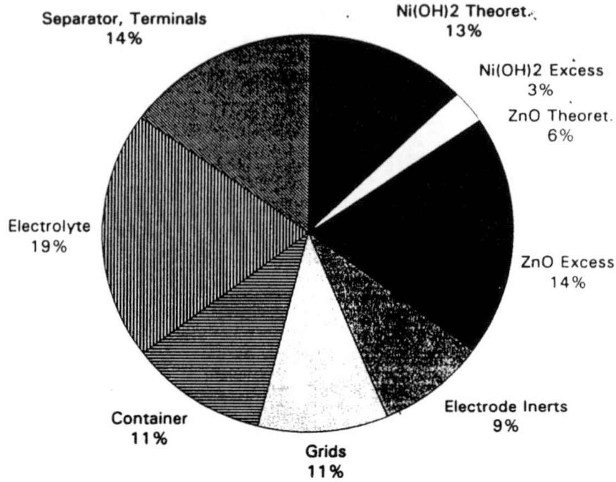

Fig. 1B Volumetric Utilization of Nickel Zinc Cell (150 AH)

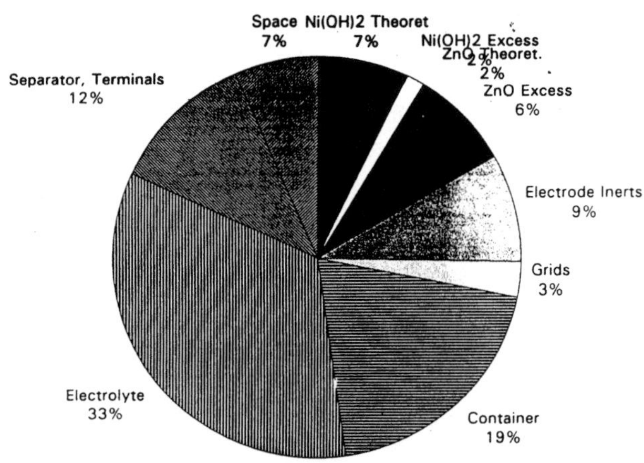

Figure 2- Fabrication of the Drumm Battery

2-8, UNION STREET,

REDDITCH.

WORCESTERSHIRE

8th August, 1966.

Dr. Alvin J. Salkind,
The Electric Storage Battery Company,
Carl. F. Norberg Research Center,
19, West College Avenue,
Yardley,
PENNSYLVANIA

My Dear Al,

I think the enclosed article from The Electrical Review gives you the best picture of the Drumm battery selected from the rubbish that was published about it.

In addition I enclose a list of other press references which we can obtain for you, but I do not think any of these contain anything which is not in the Electrical Review article.

I confirm that we made the positive plates for the few Drumm batteries that went into service, and these were of the standard Jungner type i.e. wide pockets, 23 m.m. wide and filled with thick briquettes. The electrolyte did not contain lithia and in the light of our present knowledge, apart from the silver difficulties, the positives would not have been satisfactory under the operating conditions envisaged.

I hope this may help you and if I can do anything else please let me know.

With kind regards,

Yours sincerely,

G.S. Atkinson

Encld.

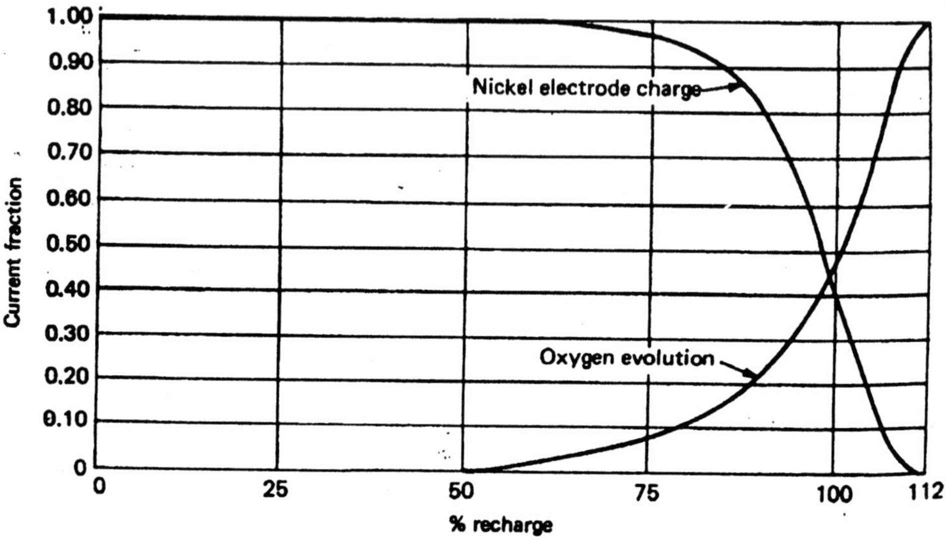

FIGURE #3 Typical nickel electrode charge acceptance.

from E. Frankowiak [ref 11]

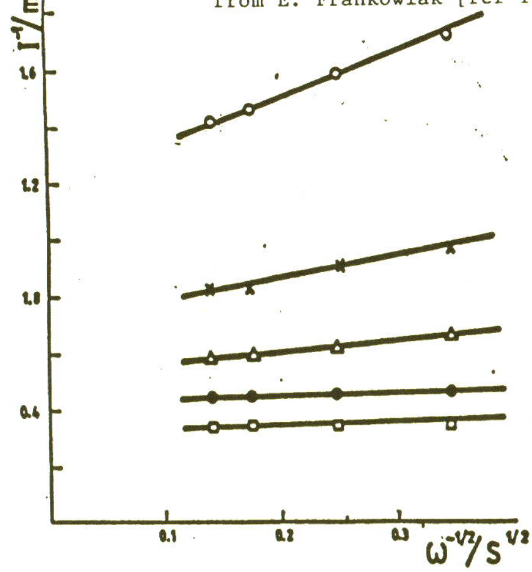

Levich plot for zinc pellet electrode without additive. Anodic overpotentials: ○ 20 mV, × 30 mV, △ 40 mV, ● 50 mV, □ 60 mV

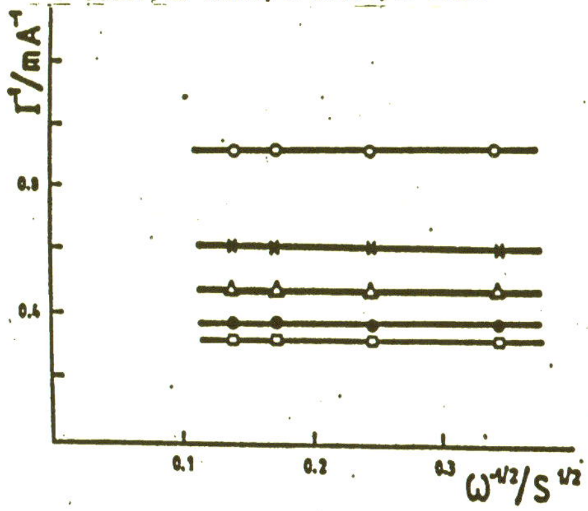

Levich plot for zinc pellet electrode with 5% AB. Anodic overpotentials: ○ 20 mV, × 30 mV, △ 40 mV, ● 50 mV, □ 60 mV

Figure 4

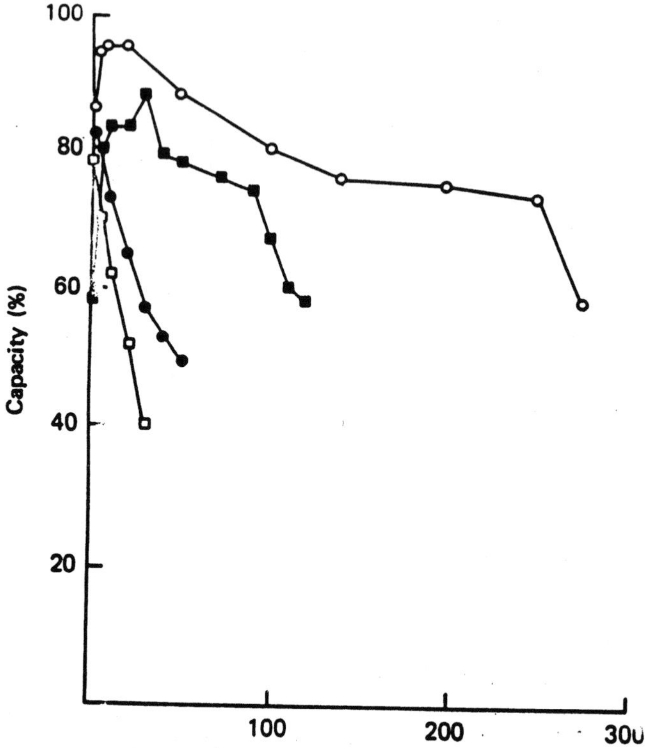

Figure 5
Capacity vs cycle number for 25 Ah cells at the 0.2 C rate to 100% DOD. graphite contents of zinc electrodes: □ 0%, ● 8%, ■ 16%, ○ 32%.

Figure 6- Effect of Rotational Field on Zinc Deposits -

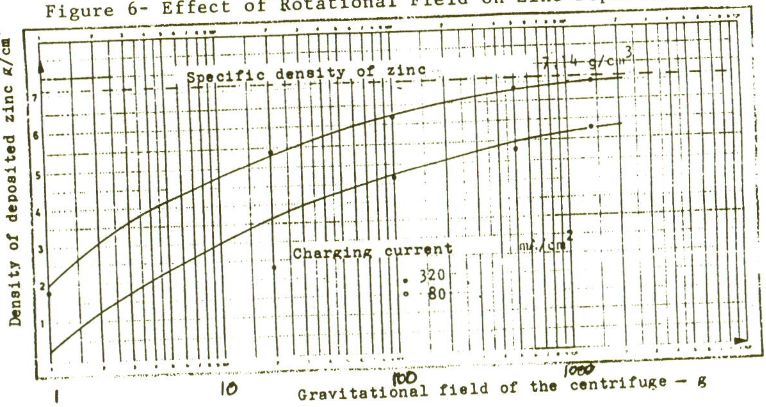

From P. Tamminen [Ref 15, 16]

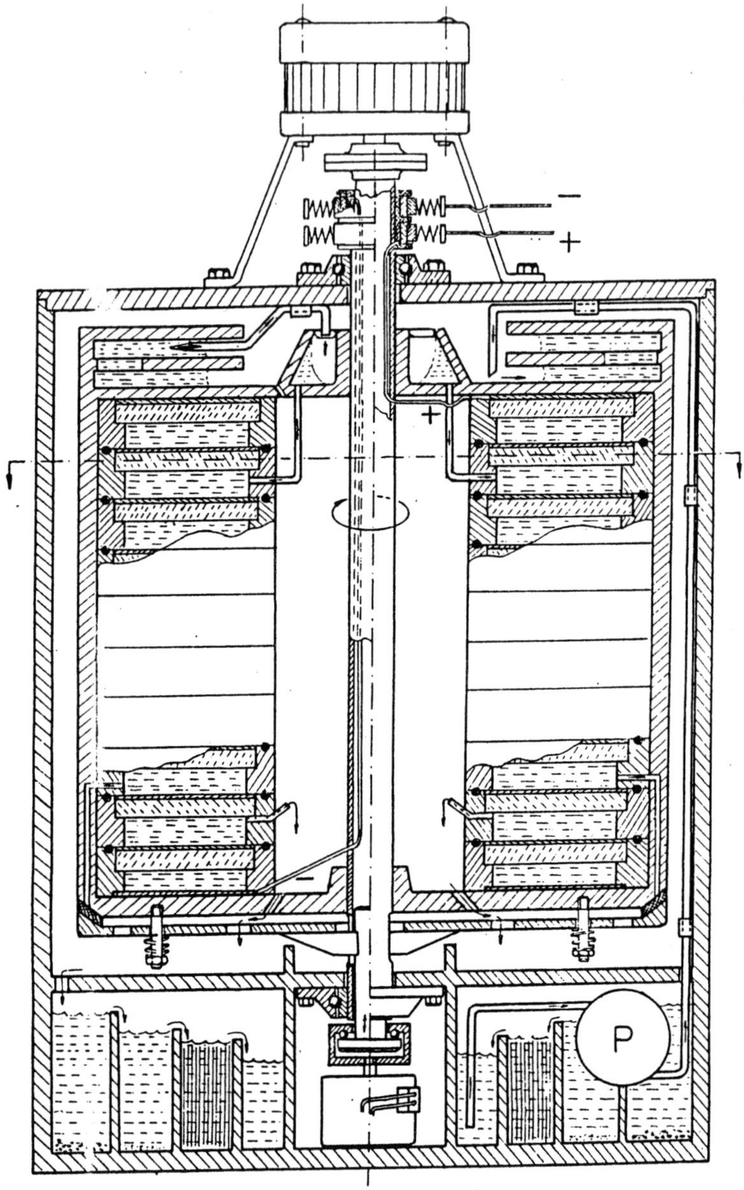

Figure 7

FIGURE [illegible] ENERGY DENSITY DEPENDANCE ON NICKEL ELECTRODE CAPACITY

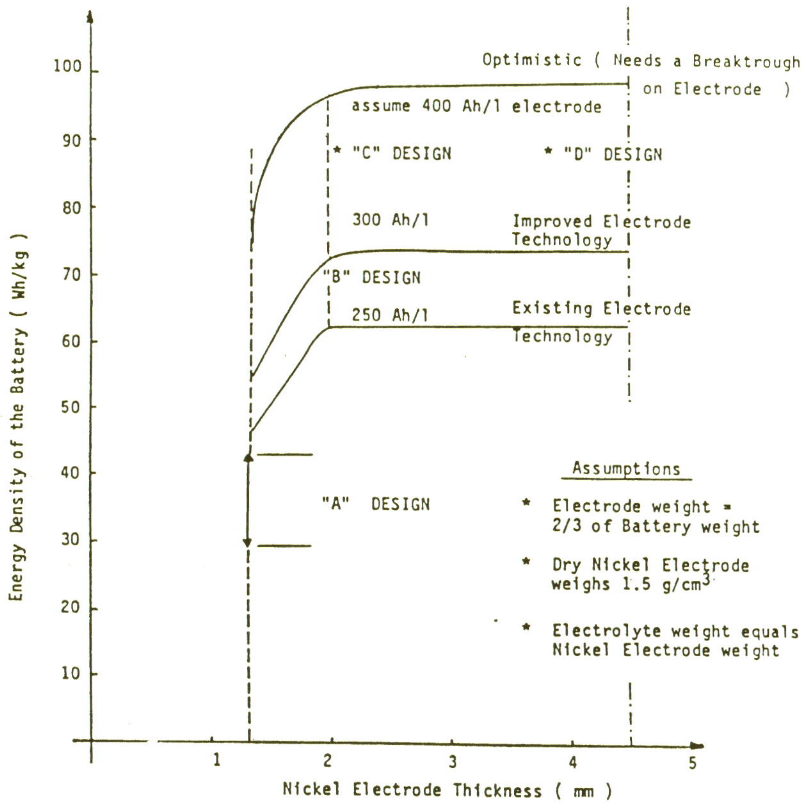

Figure 8

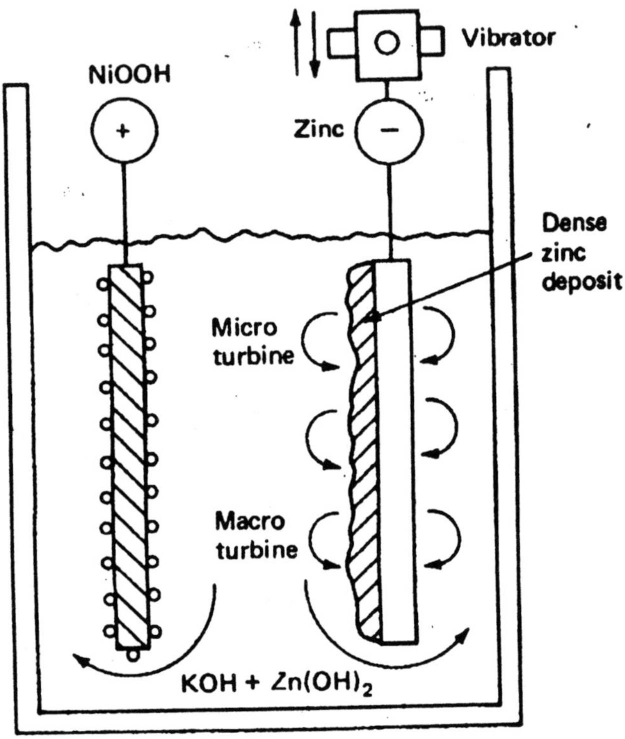

FIGURE 9 Principle of operation of nickel-zinc Vibrocel. *(Courtesy of ESB Technology Co.)*

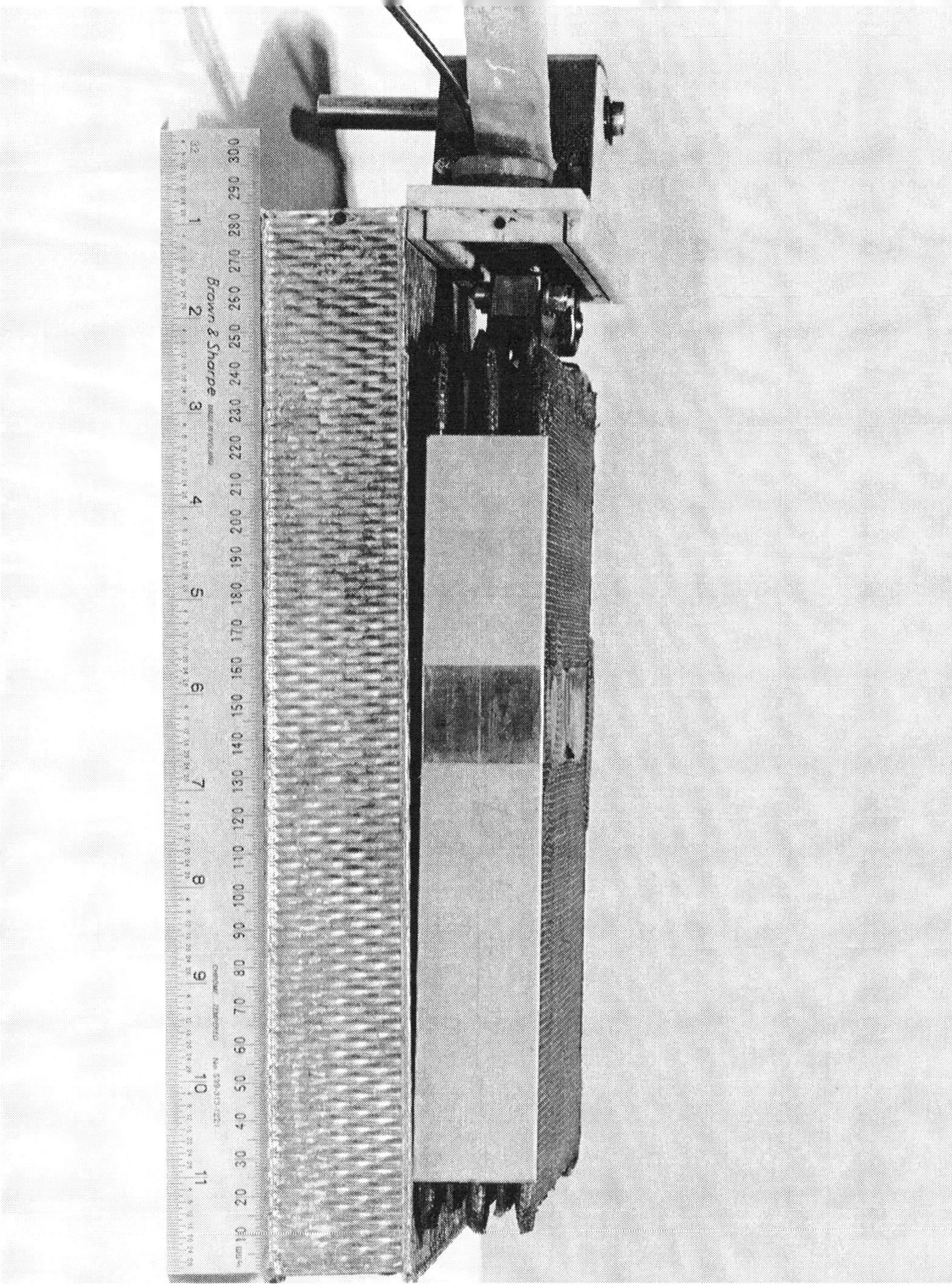
Brown & Sharpe

United States Patent [19]

Kochergiinsky et al.

[11] 3,907,603

[45] Sept. 23, 1975

[54] STORAGE CELL WITH DISSOLUBLE NEGATIVE LINE ELECTRODES

[76] Inventors: **Meer Danilovich Kocherginsky**, ul. Komarova 6, kv. 23; **Sergei Leonidovich Kalachev**, ul. Chasovaya, 26, kv. 106; **Victor Arsenievich Naumenko**, Novoslobodskaya ul., 62, kv. 64; **Lidia Fedorovna Penkova**, ul. 3 Mytischinskaya 14a, kv. 124, all of Moscow, U.S.S.R.

[22] Filed: **Aug. 15, 1973**

[21] Appl. No.: **388,641**

Related U.S. Application Data

[63] Continuation-in-part of Ser. No. 327,848, Jun. 29, 1973, abandoned, which is a continuation of Ser. No. 98,762, Dec. 16, 1970, abandoned.

[52] **U.S. Cl.** **136/140; 136/30; 136/141**

[51] **Int. Cl.²** **H01M 6/00**

[58] **Field of Search** .... 136/6 R, 30, 140, 141, 136/168, 86 A; 204/157.15

[56] **References Cited**

UNITED STATES PATENTS

383,706	4/1888	Cox	136/140
390,675	10/1888	Humphreys	136/140
2,594,713	4/1952	Andre	136/6 B
3,532,548	10/1970	Stachurski	136/86 A
3,536,535	10/1970	Lippincott	136/140
3,560,261	2/1971	Stachurski et al.	136/6 R

Primary Examiner—Patrick P. Garvin
Assistant Examiner—John F. Niebling
Attorney, Agent, or Firm—Waters, Schwartz & Nissen

[57] **ABSTRACT**

A storage cell comprising an electrolyte-filled jar in which are disposed dissoluble zinc electrodes and positive electrodes fabricated from oxides of nickel or silver or air-depolarized electrodes, with perforated separators disposed therebetween. The dissoluble zinc electrodes and positive electrodes are provided with current-collectors, the current-collectors of the dissoluble zinc electrodes being connected to a means for imparting vibration to the zinc dissoluble electrodes in a manner ensuring mechanical interaction between the whole surface of the dissoluble zinc electrodes and the plastic separators and compaction of the sponge-zinc deposit forming on the electrodes in the course of charging.

To provide intensive stirring and accelerate the dissolution of the zinc oxide deposit accumulating on the bottom of the jar in the course of discharging, the ends of the zinc electrodes adjacent to the bottom of the jar are bent at a certain angle or connected in pairs by a perforated plate.

The dissoluble zinc electrodes consist of at least two layers, the layer adjacent to the plastic separator having a lesser porosity, while the means for imparting vibration to the dissoluble zinc electrodes is a cam drive.

14 Claims, 6 Drawing Figures

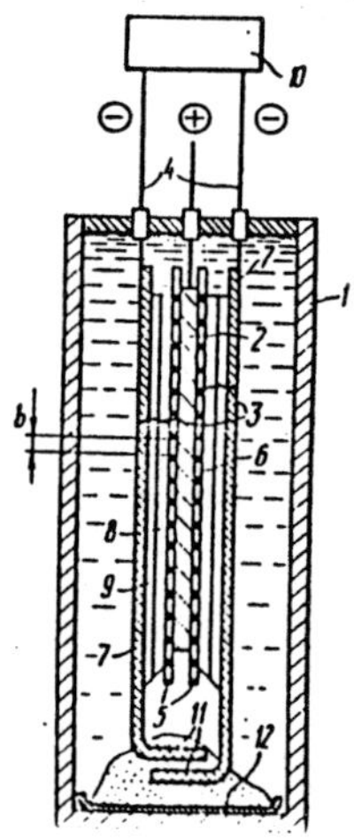

Figure 11

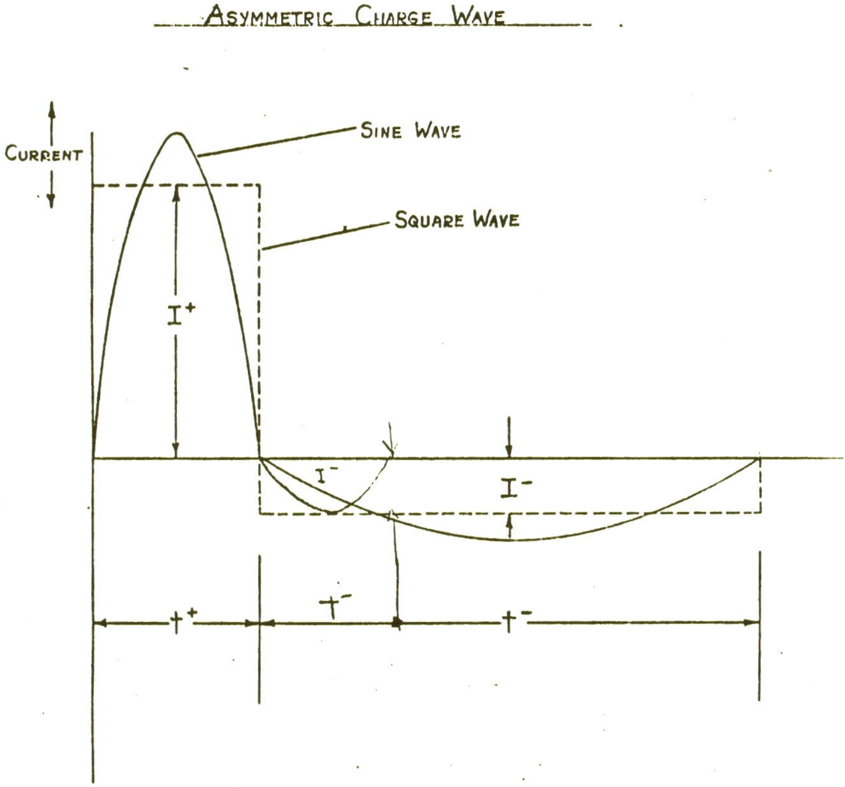

Figure 12 - [Ref. 21] C. Stoppler and C. Grun

ALKALINE SOLID POLYMER ELECTROLYTE Ni/Zn SECONDARY BATTERIES

S. Guinot, N. Bouzir, J.F. Penneau* and J.F. Fauvarque
CNAM - Laboratoire d'Electrochimie Industrielle
2, rue Conté - 75003 Paris - France
*EDF - Département des Etudes et de la Recherche
77250 Moret-sur-Loing - France

Alkaline solid polymer electrolyte nickel/zinc cells have been constructed and tested at room temperature. Several kinds of cells have been studied, depending upon the nature of the negative electrode. Cycling experiments undoubtedly show the good reversibility of all such solid polymer electrolyte cells, without any internal shorting phenomenon over the cycling period studied. The solid polymer electrolyte used here seems to be a good alternative to its liquid homologue in order to minimise the dendritic growth problem. $Ca(OH)_2$ based cells seem to exhibit the best faradaic yield values.

INTRODUCTION

History and research concerning nickel/zinc secondary batteries are one century old. The studies have been first initiated in 1887 by Dun and Haschlacher, who patented several systems based on nickel-potassium hydroxide secondary batteries (1). Then, at the end of the 19th century, both W. Jungner in Sweden and T.A. Edison in the USA developed some prototypes of nickel/zinc secondary batteries, but failed in their commercialisation, and finally gave up their studies. Since that time, the nickel/zinc secondary battery has been on one hand suffering from competition with other systems and on the other hand raising a renewed interest because of a special environmental context.

The zinc material represents a good candidate for secondary batteries, because of its numerous advantages such as a reduced cost, a relative abundance, a low toxicity, a good stability over a wide operating range and a high oxidation potential. Nickel/zinc systems exhibit good performances and interesting characteristics, as presented in table 1.

Energy density	55-85 $Wh.kg^{-1}$
Power density	140-200 $W.kg^{-1}$
Low temperature performances	very good
Reactivity of zinc for O_2 recombination	high → good candidate for sealed cells
Open circuit voltage	1.75 V
Self-discharge rate	less than 0.8% / day

Table 1: performances and characteristics of a typical Ni/Zn secondary battery (electrolyte: aqueous solution of KOH 6-7 M, from reference 2)

Despite all those qualities, nickel/zinc secondary batteries suffer from poor cycle life, because of four main complications, which arise from the zinc electrode behaviour in alkaline electrolytes (3). Those problems are:

(i) shape change
(ii) dendritic growth
(iii) zinc corrosion
(iv) zinc migration towards the nickel electrode

(i) Shape change

It takes place during cycling and corresponds to the redistribution of zinc active material over the electrode area. This phenomenon is due to the high solubility of the zinc oxidation products, the zincates $Zn(OH)_4^{2-}$, in the alkaline electrolyte. Its main consequence is a densification of zinc at the centre of the electrode which induces a loss of both the capacity and the active area, and reduces cycle life.

(ii) Dendritic growth

This phenomenon occurs during the charging process and arises from critical values of the current density and zincate concentration. Those tree-like structures grow from the negative electrode towards the positive electrode and can pierce the separator material, thus causing an internal short-circuit.

(iii) Zinc corrosion

Zinc corrosion yields zinc hydroxide and zincates, and it results in a poor charge retention of the electrode.

(iv) Zinc migration towards the nickel electrode

Chemical analysis undoubtedly prove that, during cycling, zinc migrates from the negative electrode towards the positive electrode, thus resulting in pore blocking and loss of capacity of the latter.

Present fundamental work on zinc electrode in alkaline electrolytes deals with the reduction of ZnO solubility, the effect of additives and pulse charging on the morphology of zinc, the suppression of H_2 evolution, the mechanism of formation and decomposition of supersaturated zincate electrolytes... Many studies have been carried out in order to solve all those problems, and include the use of additives for the electrode (4-8) and/or the electrolyte (9-11), the use of new separating materials (3), and changes in cell technology.

The efficiency of many additives is now established. For example, the use of oxides or hydroxides of high H_2 overvoltage metals like PbO, CdO, SnO_2, Tl_2O_3, $In(OH)_3$, Ga_2O_3 or mixtures of them, is preferred to HgO to suppress H_2 evolution. Actually, HgO is known to accelerate the rate of shape change (12).

The effectiveness of $Ca(OH)_2$ to reduce shape change has been extensively studied and demonstrated (6). However, current studies tend to show the beneficial effect of Bi_2O_3, which influences crystal growth from the electrode surface, thus avoiding the problem of zinc electrode passivation caused by $Ca(OH)_2$ (insoluble discharge species blocking the porous electrode surface) (13).

In the field of the present research on alkaline Ni/Zn secondary batteries, we present here a new approach consisting in the replacement of the conventional aqueous electrolyte by a solid polymer electrolyte (SPE).

First proposed by Armand in 1978, the concept of solid polymer electrolytes for lithium secondary batteries has led to a tremendous research effort during the last 20 years (14). Many advantages such as chemical and electrochemical stability, good mechanical properties, ease of handling, dual role of mechanical separator and ion conducting medium and flexible geometry make such electrolytes attractive for their use in secondary batteries.

Five years ago, our laboratory initiated the study of a new class of SPE which specificity lies in their composition based on poly(ethylene oxide) PEO, potassium hydroxide KOH and water (15-16). In addition to a structural study, we examined the opportunity of using such electrolytes in nickel/zinc secondary batteries (17-18). The aim of our study is based on the assumption that zinc oxidation products may be less soluble in our SPEs than in the corresponding liquid alkaline electrolyte. Thus, such solid medium might impede dendritic growth and reduce or avoid shape change, as there is no or less convection inside. We expect to obtain a second generation zinc electrode with enhanced cyclability.

We have studied and characterised numerous SPE compositions in the following range:

- PEO: 40-80 wt. %
- KOH: 8-50 wt. %
- H_2O: 0-20 wt. %

Our most studied composition 60% PEO/30% KOH/10% H_2O has been chosen for the construction of the cells. We recall here, as a comparison, the main results concerning the study of the first Ni/SPE/Zn secondary battery built in our laboratory (capacity: 80 mAh), which preliminary results have been presented elsewhere (12).
The cycling results of four 640 mAh Ni/SPE/Zn cells with different zinc electrode compositions will also be discussed.

EXPERIMENTALS

SPE preparation

Commercial PEO (Fluka, av. mol. wt. 20,000) and potassium hydroxide (Fluka) were used as starting materials. The 60% PEO/30% KOH/10% H_2O composition was prepared by direct mixing of the components at approximately 90°C - 100 °C, and was allowed to cool slowly at room temperature.

Cell design

80 mAh cell. The negative electrode is an expanded zinc grid, welded on each side on a cadmium drilled sheet as a collector. The positive electrodes are made of sintered nickel, and have a theoretical capacity of 160 $Ah.m^{-2}$. The cell was manufactured by assembling a

central circular zinc electrode (thickness: 2.10^{-4} m, area: 5.10^{-4} m^2), between two nickel electrodes (thickness: 10^{-4} m, area: 5.10^{-4} m^2), each electrode being coated with a SPE layer (figure 1).

The cell capacity is 80 mAh. It is limited by the positive electrodes. The zinc electrode capacity value is almost five times higher, i.e. 820 $Ah.m^{-2}$.

640 mAh cells. The negative electrode (thickness: 10^{-3} m, area: 15.9 10^{-4} m^2) is a ZnO-based paste of various compositions containing ZnO (Prolabo), CdO (Wiaux), methylcellulose (Fluka), PTFE (Dupont de Nemours), $Ca(OH)_2$ (Aldrich), C (Cabot) and PbO (Prolabo), and deposited on a cadmium coated stainless steel collector. The positive electrode (thickness: 10^{-3} m, area: 15.9 10^{-4} m^2) is a nickel foam filled with $Ni(OH)_2$ active material, and has a theoretical capacity of 400 $Ah.m^{-2}$. Figure 2 presents the design of the negative and the positive electrodes. Both nickel and zinc electrodes were previously impregnated with a little amount of aqueous solution of KOH, then covered with a SPE powder layer and kept in an oven at 80°C for one hour. The separator we call "SPE membrane" was prepared separately almost in the same manner, but with no previous liquid electrolyte impregnation. The last operation was the cell assembling (see figure 2) and sealing in a polyethylene bag.

The cell capacity is 640 mAh. It is limited by the positive electrode. The zinc electrode capacity ranges from 2.4 to 3 times this value, depending on the composition.

Cell testing

All tests were performed at room temperature.

80 mAh cell. Cell testing was performed using a CEAMID-6 type chrono-amperostat. Cycling experiments were made under the following conditions:

- potential range: 1 V - 2 V
- charge rate: C/16
- discharge rate: C/8

640 mAh cells. A Mac-Pile type computer-controlled cycler was used to perform constant current charge and discharge of the cells. Cycling experiments were carried out under the following conditions:

- potential range: 1.4 V - 1.975 V to 2.2 V
- charge rate: C/26
- discharge rate: C/13

RESULTS AND DISCUSSION

The reactions involved in the charge-discharge process of the Ni/Zn elements are the following:

$$Ni(OH)_2 + OH^- \underset{discharge}{\overset{charge}{\rightleftarrows}} NiOOH + e^- + H_2O \quad [1]$$

$$ZnO + 2e^- + H_2O \underset{discharge}{\overset{charge}{\rightleftarrows}} Zn + 2OH^- \quad [2]$$

The overall reaction being:

$$Zn + 2NiOOH + H_2O \leftrightarrow ZnO + 2Ni(OH)_2 \quad [3]$$

The zinc discharge process is based on both an electrochemical and a dissolution-precipitation reaction:

$$\begin{aligned} &Zn + 4OH^- \rightarrow Zn(OH)_4^{2-} + 2e^- \\ &Zn(OH)_4^{2-} \rightarrow ZnO + 2OH^- + H_2O \end{aligned} \quad [4]$$

80 mAh cell

The specificity of this element is the fact that its initial state is a charged state, which means that the starting material is metallic zinc. This element exhibits a good discharge behaviour, even under a rate of C/8. Its internal resistance is 6 Ω, which is a reasonable value for such SPE cell. Figure 3 shows typical charge-discharge curves for this Ni/Zn cell, which prove its good reversibility. This element has been cycling for 60 cycles without any short-circuit and dendritic phenomenon, before failing from total zinc consumption (figure 4). Faradaic yield values up to 87% could be obtained. However, a decrease of capacity has been observed after the 47th cycle. Table 2 summarises the main characteristics of the cell.

Nominal voltage at charged state	1.85 V
Nominal voltage at discharged state	1.52 V
Energy density	112 Wh/kg*
Specific capacity	70 Ah/kg
Energetic yield	70 %
Cyclability	60 cycles
Internal resistance	6 Ω

Table 2: performances of the 80 mAh SPE Ni/Zn cell
(* active materials only)

In order to better understand this faradaic yield problem, we have built another kind of cell, with zinc oxide as starting material, and a capacity value of 640 mAh.

640 mAh cells

Four experimental cells were constructed and studied. Table 3 lists the nature and the quantity of materials used to prepare the negative electrode of these cells. The difference between cell n°2 and cell n°3 lies in the preparation mode of the pasted zinc electrode. Indeed, cell n°3 negative electrode components have been previously blended in a grind to produce a so-called "homogeneous paste".

Zinc electrode composition (wt. %)	Cell 1	Cell 2	Cell 3 (homogeneous paste)	Cell 4
ZnO	69.4	76.53	76.53	66.3
CdO	11.22	12.25	12.25	13.3
Methocel	1.02	1.02	1.02	1.02
PTFE	3.06	3.06	3.06	3.12
$Ca(OH)_2$	15.3	-	-	13.3
C	-	5.1	5.1	1.63
PbO	-	2.04	2.04	1.33

Table 3: zinc electrode composition of the four 640 mAh Ni/SPE/Zn cells

<u>Cell forming</u>. Throughout testing, the four cells have been subjected to changes in the cycling conditions, before working under almost similar conditions for all cells. This preliminary testing or "formation period" provided much information concerning the cycling behaviour of the cells in the solid polymer electrolyte medium, and allowed to determine appropriate cycling conditions which are summarised in table 4. The formation period is difficult, because of the SPE medium. It also appears to be very long and very important as well. Indeed, in such "all solid cell", the wetting process of the electrodes by the SPE is progressive, and usually improves with cycling. The four cells have been subjected to different cycling conditions during the course of this formation period, as shown by the erratic behaviour of the charged capacity displayed in figure 5.

In order to avoid an overcharge of the cells, the cycling process has to be progressive, by charging on average 1.5 to 2 times the equivalent amount of the discharged capacity obtained during the previous cycle. Thus, the appropriate potential range for a cycle may be 1.4 V-1.975 V to 2.2 V (depending upon cell composition), to ensure the charge and the discharge of the zinc active material only, in the negative electrode. The lower limit (1.4 V) corresponds to the end of discharge potential of zinc. The upper limit 1.975 V to 2.2 V is a compromise between two phenomena observed at the following potential values:

- 1.8 V, which is the minimal value to exceed in order to ensure the charge of ZnO ;
- 1.975 V to 2.2 V, the value above which O_2 evolution can occur and damage the cell.

Cell n°	Potential range (V)	Charge rate	Discharge rate
1	1.4 - 1.975	C/26	C/13
2	1.4 - 1.975	C/26	C/13
3	1.4 - 2.2	C/26	C/13
4	1.4 - 2.2	C/26	C/13

Table 4: cycling conditions for the 640 mAh cells

It seems difficult to work on more than a certain proportion of the theoretical capacity, at least at the beginning of the cell life (see table 5).

Cell number	Test/formation period (cycle number)	Maximum value of discharge capacity (% theoretical capacity)
1	41	30
2	28	40
3	36	20
4	70	30

Table 5: discharge capacity during the formation period

Cycling experiments. As an example, figure 5 shows the cycling behaviour of cell n°1. The arrow on the graph represents the cycle number from which the cell has been tested under the conditions summarised in table 4.

All experiments show that the faradaic efficiency value decreases with increasing charge capacity, at the working charge current, i.e. 25 mA. These results indicate the existence of a self-discharge phenomenon. This is confirmed by further experiments, which show that the use of a greater charge current value allows to enhance significantly the faradaic yield. Studies are under progress to identify the reasons of this self-discharge phenomenon.

Figure 6 shows smoothed curves of the faradaic yield evolution versus cycle number for all four cells, after the formation period. Over a common cycling period, the faradaic values can be classified in the following manner:

cell 1 > cell 2 > cell 4 > cell 3.

One can observe the cell n°1 and cell n°2 values are very close to each other, as well as cell n°3 and cell n°4 values. With a good zinc electrode composition, good values of faradaic yield can be obtained in the range of 40% to 80%, as it is the case for cell n°1, the negative electrode of which contains $Ca(OH)_2$. It seems that for the two identical compositions (cell n°2 and cell n°3), the homogenisation of the zinc electrode components has no effect on faradaic efficiency. In the same way, mixing together all the different components that are used separately for cells n° 1, 2 and 3 (cell n°4), provides the lowest values. Figure 7 shows typical charge and discharge curves for cell n° 1, as an example. The discharge curve exhibits a very clear end of discharge signal. However, the end of charge has not been reached yet, under our cycling conditions. This phenomenon can be observed for all cells.

All four cells are still currently under cycling, with increasing charge rate values. Indeed, they suffer from a self-discharge phenomenon, and the charge current value has to be greater, both to exceed the zinc self-discharge current value and to reduce the duration of a cycle. Increasing values of the faradaic efficiency are obtained (see figure 8).

It is also important to notice that similar experiments carried out in an aqueous alkaline electrolyte have shown dendritic growth before approximately five cycles.

CONCLUSION

Despite intermediate values of the faradaic efficiency, it is possible to perform many cycles with a solid polymer electrolyte nickel/zinc secondary battery, both with zinc or zinc oxide as starting materials, and without any dendritic growth and shape change. This undoubtedly proves that the zinc electrode can de charged and discharged in our solid polymer electrolyte medium. Once the electrodes are formed, it is possible to increase the depth of charge and the current densities. Such experiments are currently under progress, and increasing values of the faradaic yield have already been obtained.

The main problem of such secondary batteries remains a significant self-discharge phenomenon. The goal of our studies is now to identify the main reasons for this phenomenon. For example, it would be interesting to elaborate and test other cells, with a negative electrode containing impurities such as iron. It would also be important to quantify the corrosion current value of zinc in such SPE medium.

ACKNOWLEDGMENTS

The authors thank The Agence de l'Environnement et de la Maîtrise de l'Energie and Electricité de France for financial support.

REFERENCES

1. A. Dun and F. Haslacher, German Patent 38, 383 (1887).
2. A. Duffield in Modern Battery Technology, C.D.S. Tuck, Editor, p. 452, Ellis Horwood, Chichester (1991).
3. F.R. McLarnon and E.J. Cairns, J. Electrochem. Soc., **138**, 645 (1991).
4. J. McBreen and E. Gannon, Electrochimica Acta, **26**, 1439 (1981).
5. J. McBreen and E. Gannon, J. Electrochem. Soc., **130**, 1980 (1983).
6. E.G. Gagnon and Y.M. Wang, J. Electrochem. Soc., , 2091 (1987).
7. E.G. Gagnon and B.S. Hill, J. Electrochem. Soc., **137**, 377 (1990).
8. R. Jain, T.C. Adler, F.R. McLarnon and E.J. Cairns, J. Appl. Electrochem., **22**, 1039 (1992).
9. R.F. Thornton and E.J. Carlson, J. Electrochem. Soc., 127, 1448 (1980).
10. J.T. Nichols, F.R. McLarnon and E.J. Cairns, Chem. Eng. Commun., **37**, 355 (1985).
11. T.C. Adler, F.R. McLarnon and E.J. Cairns, J. Electrochem. Soc., **140**, 289 (1993).
12. J. McBreen, J. Electrochem. Soc., **119**, 1620 (1972).
13. J. McBreen and E. Gannon, J. Power Sources, **15**, 261 (1985).
14. M.B. Armand and J.M. Chabagno in Fast Ion transport in Solids, P. Vashishta, J.N. Mundy and G.K. Shenoy, Editors, p. 131, North Holland, New-York (1978).
15. N. Bouzir: Ph.D. Thesis - Université Pierre et Marie Curie - Paris - 1994.
16. J.F. Fauvarque: European Patent Application 93919409.8.
17. J.F. Fauvarque, S. Guinot, N. Bouzir, E. Salmon and J.F. Penneau, Electrochimica Acta, **40**, 2449 (1995).
18. S. Guinot: Ph.D. Thesis - Université Pierre et Marie Curie - Paris - 1996 (to be presented).

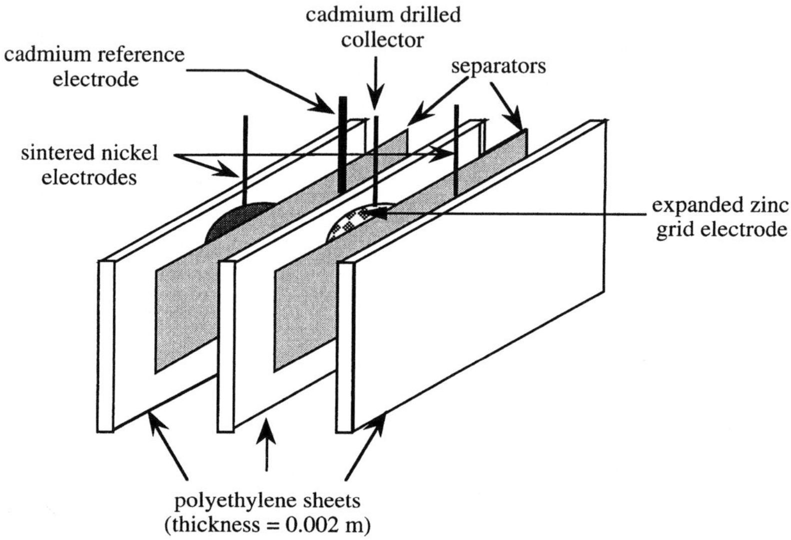

Figure 1: 80 mAh solid polymer electrolyte Ni/Zn cell composition.

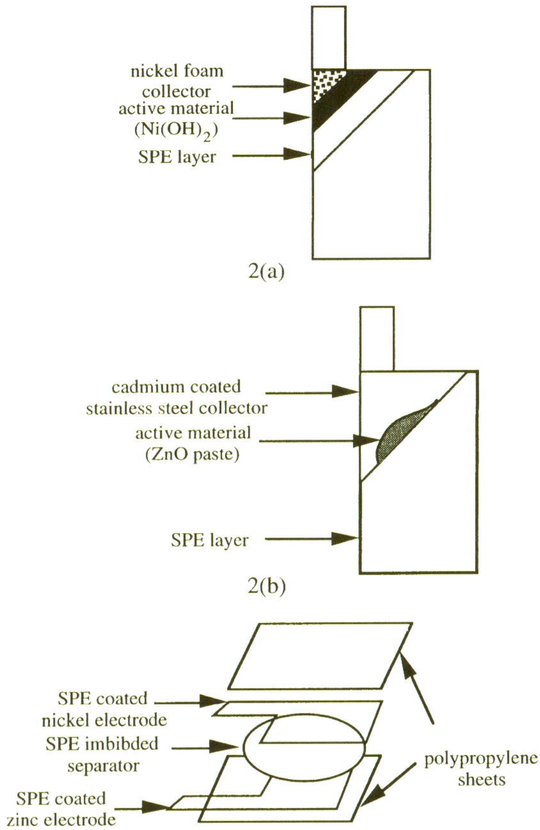

Figure 2: typical Ni/Zn cell construction.

(a) nickel electrode construction
(b) zinc electrode construction
(c) cell assembling

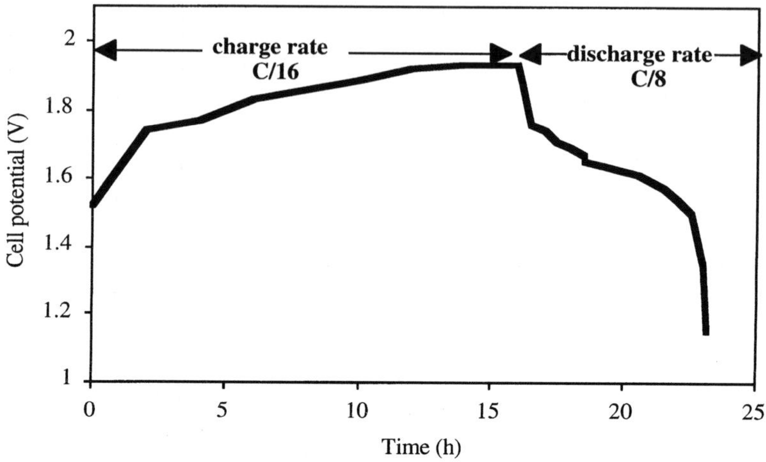

Figure 3: typical charge-discharge curves for a 80 mAh SPE Ni/Zn cell.

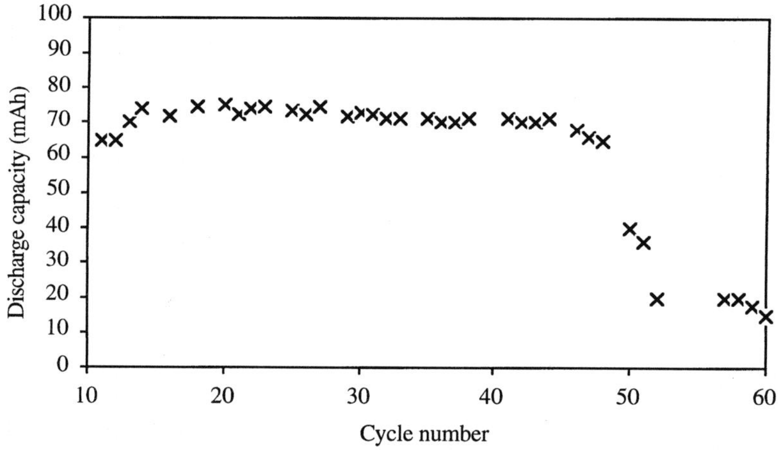

Figure 4: discharge capacity versus cycle number for the 80 mAh SPE Ni/Zn cell.

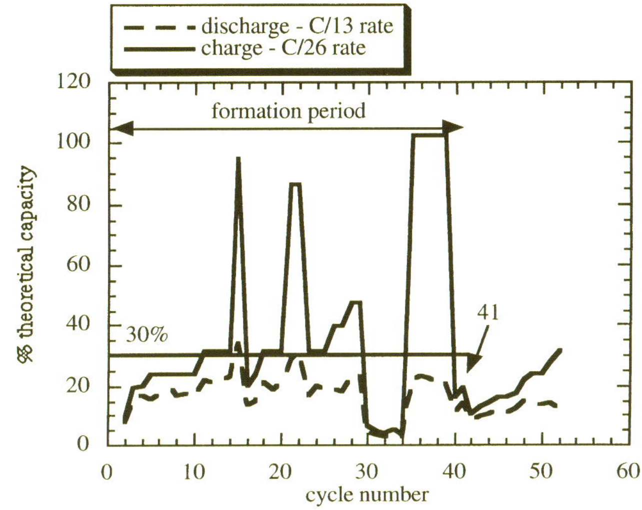

Figure 5: charge and discharge capacity versus cycle number - 640 mAh cell (cell n°1).

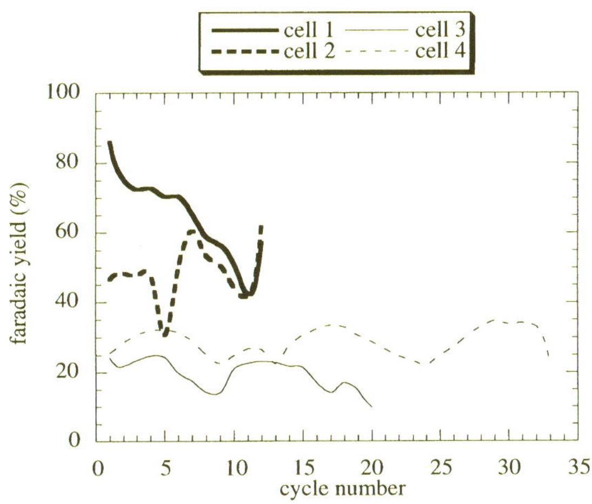

Figure 6: variation of the faradaic yield versus cycle number for the four 640 mAh cells.

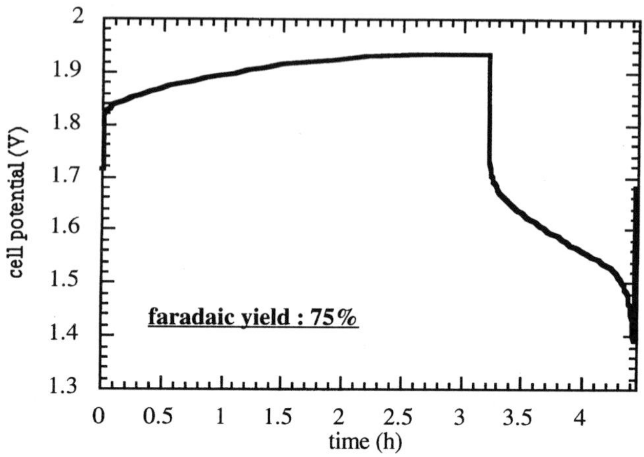

Figure 7: typical charge-discharge curves for a 640 mAh Ni/Zn cell (charge rate: C/26 - discharge rate: C/13).

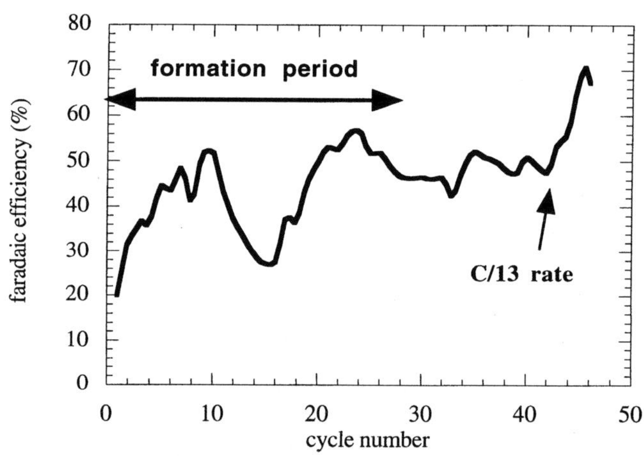

Figure 8: variation of the faradaic efficiency versus cycle number for cell n°1

IMPEDANCE STUDIES OF ROLL-PRESSED POROUS ZINC ELECTRODE FOR Ni-Zn BATTERY

S.G. Doo, H.B. Lee and J.Y. Oh

Energy & Environment Laboratory, Materials & Devices Research Center, Samsung Advanced Institute of Technology, P.O. Box 111, Suwon, Korea

ABSTRACT

The purpose of this work is to suggest the appropriate equivalent circuit model for better understandings of the complex impedance behavior of porous zinc electrode in various experimental conditions. All impedance data could not be explained by ideal equivalent circuit which is cited in the literature because of its complexcity. Davidson-Cole model considering the roughness and porosity of electrode(i.e. roughness fractality) was introduced instead of simple RC circuit. Moreover, impedance data measured for the electrode of low state of charge were very different from that of high SOC, regardless of experimental conditions. Two types of equivalent circuit model were suggested to explain the impedance behavior at all experimental conditions. Model I which has a good agreement with the electrode of low SOC, contains three frequency domains; ZnO characteristics(HF), the corrosion reaction and double layer behavior(IF), and the diffusion process(LF). Model II for the electrode of high SOC, contains the corrosion reaction and double layer behavior(HF) and adsorption process (LF).

INTRODUCTION

Zinc electrode for the Ni-Zn battery has several advantages i.e. electrochemical reversibility in alkaline electrolytes and relatively high energy density etc., but it has some problems which limits the cycle life, such as shape change, densification, passivation and dendrite formation. Recently many investigations have been performed to enhance the cycle life of zinc electrode(1-4).

First of all, it is necessary to understand the phenomena occured in the electrode interface for the enhancement of the battery performance. Electrochemical impedance spectroscopy(EIS) has extensively used for the study of the reaction mechanism(5). The applications of impedance techniques for characterization of battery have recently been supposed because it is somtimes a good diagonostics for evaluating the state of batteries(6).

Cachet and Wiart(7) investigated the pore texture of zinc electrode characterized by impedance measurements and cylindrical pore model(8), but the manufacturing method of the tested electrodes were different from that of the commercial electrode. Some impedance investigations for porous zinc electrode have recently been carried out on flooded system(9) and starved elctrolytes condition(10) to understand its electrochemical behaviors, but the suggested equivalent circuit models are unsatisfactory to explain the complex impedance behaviors abtained in this study. Moreover, it was reported that the impedance techniques appeared not to be suitable for monitoring the special information, such as shape change, from the impedance spectra measured while cycling(11).

Therefore the objective of this work is to suggest the satisfactory equivalent circuit models for better understandings of the complex impedance behavior of porous zinc electrode in various experimental conditions.

EXPERIMENTAL

Roll-pressed porous zinc electrodes(theoretical capacity: 1Ah, size: 2cm × 3.27cm) were prepared with dry state mixtures of 87wt.% ZnO, 7wt.% Zn powder, 2wt.% CdO, 1wt.% PbO and polytetrafluoroethylene(PTFE) and a Cd-coated Cu mesh was used as a current collector. A sintered nickel oxide electrode(theoretical capacity: 1Ah, size: 6 cm × 4.6 cm) taken from commercial Ni-Zn batteries serves as a counter electrode and one layer of Celgard 3401 as a separator. Hg/HgO reference electrode was connected to test cell with three electrodes system through a capillary and sufficient elctrolyte volume of 8.6 M KOH and 1 M LiOH was added because working electrode and counter electrode have different surface area. A precise description of the test cell has been given elsewhere(12). All potentials are expressed against Hg/HgO electrode. The state of charge(SOC) is defined as the theoretical charging amount, i.e. the charging efficiency is not considered. Galvanostatic method was taken to study the charging characteristics of the electrode.

The impedance measurements were made using M398 system(EG&G Princeton Applied Research) incorporated a frequency response analyser(M5208 Lock-In Amplifier), a electrochemical interface(M273A Potentiostat/Galvanostat) and a personal computer, and a schematic diagram of experimental setup is shown in Fig. 1. Impedance measurements were made at the frequencies in the range of 2 mHz to 100 kHz and the amplitude of AC potential was 5 mV. All impedance measurements were carried out at open circuit potential after 2 hr idling to attain a steady electrode potential as a function of SOC, charge rate, discharge rate, cycle number.

RESULTS AND DISCUSSION

In general appropriate equivalent circuit model which shows the same impedance characteristics as experimental data is need to understand the phenomena occured in the electrode interface and examine the reaction mechnism closely by using EIS. All impedance data performed in this study showed the different impedance behavior according to the amount of charging and could not be explained by one ideal equivalent circuit model which is cited in tne literature because of its complexcity.

Equivalent Circuit Model for the Zinc Electrode Having Relatively Low SOC (Model I)

To model the impedance characteristics of the studied electrode, it was assumed that the surface of Zn metal is partially covered with ZnO in the Zn electrode having relatively low SOC. It can be seen that this is similar to a metal/organic coating/solution interface system because the corrosion reaction may proceed both on the metal surface and via the ZnO pores penertrated with electrolyte. Recently these insigts were concretized into the equivalent circuit model shown in Fig. 2 and applied to the porous zinc electrode operating under starved electrolyte conditions by Jindra and Becker-Rose[10]. The symbols used in this circuit have the following meanings.

R1: the resistance of the electrolyte between the zinc electrode and the reference electrode
R2: the resistance of the conducting pores in the ZnO layer
R3: the charge transfer resistance of the corrosion reaction occuring at the active sites on the Zn metal
C1: the capacitance on the ZnO layer
C2: the double layer capacitance at the zinc metal/electrolyte interface
W: the Warburg diffusion impedance

The impedance behavior of this equivalent circuit shows three distinguished parts (two semicircles and one straightline which makes an angle 45 degrees with both axis) in the Nyquist plot, indicated with HF(high frequencies), IF(intermediate frequencies) and LF(low frequencies). Impedance behavior of ZnO layer which, in addition to semiconducting properties, also has conducting properties via pores which are penertrated with electrolytes is characterized by semicircle at HF. This model presents a good clue to uderstand the comlex reaction mechanism, but it was unable to fit the measured impedance data well because in reality two depressed semicircle at HF and IF domain was observed and moreover the slope of straight line at LF domain deviated from 45 degrees. It was therefore thought that it was unsatisfactoray to apply the model described above to this system directly to obtain the optimum value of circuit element from the measured data by nonlinear fitting method. It has been recognized that a deforming or flattening effect(i.e. frequency dispersion) on the impedance plot was ascribed to the roughness and porosity of electrodes(14), which have an impedance Z_a as follows, whether or not sustaing charge transfer at their interface:

$$Z_a \sim z_a^{\alpha} \qquad [1]$$

is called Davidson-Cole model(15), where z_a is the microscopic impedance per unit real surface area of the double layer interface and α, a frequency independent proportionality factor, is a fractional exponent, with $0 < \alpha < 1$. It was therfore decided to apply Davidson-Cole model considering the roughness fractality of the electrodes instead of simple RC circuit.

<u>Equivalent Circuit Model for the Zinc Electrode Having Relatively High SOC (Model II)</u>

Armstrong and Bell[16] have suggested that the active dissolusion of zinc in alkaline solution proceeded with a two step mechanism, which include the adsorbed intermediate, using both steady-state potentiostatic techniques and ac impedance measurements. One of the equivalent circuit related to the adsorption process is shown in Fig.3(17) and the symbols used in this circuit have the following meanings.

R_s : the resistance of the electrolyte between the zinc electrode and the reference electrode

R_{ct} : the charge transfer resistance of the corrosion reaction occuring at the active sites on the Zn metal

R_a : the adsorption resistance of Zn intermediates

C_{dl} : the double layer capacitance at the zinc metal/electrolyte interface

C_a : the adsorption capacitance of Zn intermediates

The impedance behavior of this equvalent circuit shows two semicircles in the Nyquist plot, indicated with HF and LF. The Semicircle at HF indicates the behavior of Zn metal/electrolyte interface and semicircle at LF domin reflects the effect of adsorption/desorption process of Zn intermediates. But the experimental impedance data performed in this condition showed two depressed semicircles. Theoretical Impedance equation calculated from the above equivalent circuit considering the roughness fractality of the elctrodes was used to obtain the optimum value of elcctrical elements.

The Analysis of the Measured Impedance Data

The suggested equivalent circuits considering the roughness fractality were transformed mathmetically to transfer functions and the optimum values of electrical elements in the equivalent circuits were calculated so that the total sum of absolute difference between the measured data and the calculated value at each frequencies minimize by nonlinear least square method such as Marquardt algorithm[18].

Inductive tail at frequencies enough high in the measured impedance data(Nyquist plot) was observed in all experimental contition. This phenomena can be seen only in the electrode having porous structure, but not in mocroelectrode system. Gutmann[19] explained this by introducing pseudo-inductance, and Lazarides and Hampson[20] concluded that inductive behavior in the impedance spectra of porous lead dioxide microelectrodes was due to the distributed oxygen evolution reaction. It thus seems that inductive tail in our system is ascribed to hydrogen evolution occured in the pore. This phenomena could be simulated by the equivalent circuit which incorporates a external, parallel inductor-resistor circuit, but not considered in the modelling of the measured impedance spectra.

As mentioned above, the dissolution of Zn metal and the hydrogen evolution reaction proceed simultaneously on the Zn metal surface by corrosion at open circuit potential(corrosion potential). The anodic dissolution mechanisms of Zn were suggested by many reserchers[21-25] and the overall disolution-precipitation reaction has been shown to be

$$Zn + 4OH^- \rightarrow Zn(OH)_4^{2-} + 2e^- \quad [2]$$

$$Zn(OH)_4^{2-} \rightarrow ZnO + H2O + 4OH^- \quad [3]$$

The anodic behavior of Zn electrode in alkaline solution could be divided into active dissolution , linear dissolution, prepassive and passive region as anodic overvoltage increases[26]. It is thus expected that the dissolution of Zn electrode at open circuit potential in KOH solution proceeds in active dissolution region because of its relatively low anodic overvoltage.

The prepared Zn electrode was deeply discharged gavanostatically at 5 mA more than the theoretical capacity of Zn metal added to electrode as conducting material until Zn electrode potential leads to 1.0 V, to compare the impedance spectra of deep discharged electrode with that of current collector immersed into electrolyte and prepared electrode. The impedance data for deeply discharged elcetrode showed much larger semicircle than that for initilly prepared electrode because the amounts of Zn metal in the electrode decreases after discharge ,but smaller semicircle than that for current collector, as shown in Fig. 4. It can be seen that Zn metal in the prepared electrode is not oxidized to ZnO by fully deep discharge.

The prepared electrodes were charged galvanostatically at 40 mA to make 10%, 20%, 36%, 60% and 100% of the theoretical capacity of the zinc active material. The impedance measurements were made at the open circuit potential after charging and the effect of SOC on the impedance spectra is shown in Fig. 5. Model I was used to simulate the impedance data of the electrode in the 10 %,20% and 36% SOC conditions. In case of the electrode in the 60%, 100% SOC condition, Model II was used because Model I did not agree well reasonably with the impedance data. The calculated impedance spectra and the optimum element value are given in Fig.5 and Table I. It can be seen that the fitting reslts are excellent. It seems that the dissolutiom mechanism of the prepared Zn electrode changes according to the SOC of the Zn electrode. As shown in Table I, in case of the Zn electrode having relatively low SOC, when the SOC increases, the value of the charge transfer resistance, R3 , decreases due to the increase of the amount of Zn metal, while the value of C2 increases. On the other hand, in case of the Zn electrode having relatively high SOC, when the SOC increases, the value of R_{ct} decreases due to the increase of the amount of Zn metal, while the value of C_{dl} increases, but the values related to adsorption were nearly constant.

Fig. 6 and Table II show the effect of the discharge rates on the impedance spectra of the Zn electrode and the fitting results by Model I. The impedance measurements were carried out after the electrodes in the 36% SOC condition is discharged galvanostatically at 0.08 A, 0.2 A and 0.4 A until the cell voltage leads to 1.2V respectively. The good agreements were attained for 0.2A and 0.4A, but there are some deviations at low frequencies for the electrode discharged at 0.08A. As shown in Table II, it can be seen that the increasee of the discharge rates results in the decrease of the values of R3 and Warburg coefficient, but the values of C1 shows a reverse tendency. This results can be explained as follows. The more amount of Zn metal remains in the electrode discharged at larger current compared with that discharged at a smaller current due to smaller discharge capacity. Moreover, it seems that the more amount of ZnO is precipitated in the electrode discharged at a larger current because the concentration of zincate ion is saturated within a faster time during discharge. When the electrode is discharged at a larger current, the more amount of zincate ion remains in the electrode surface, and the diffusion of hydroxide ion to electrode surface is much more hindered by the existence of more zincate ion and the dense ZnO layer.

The cycle test of the test cell was performed to investigate the variation of discharge capacity, cell voltage and Zn electrode potential versus cycle number and the charge efficiency(the percent of discharge capacity/charge capacity) versus cycle number during the cycle test were shown in Fig. 7. The test cells were charged galvanostatically at 40mA with 36% SOC, which corresponded to 90% charging for the commercial Ni-Zn battery and discharged at 80 mA until the cell voltage reached the discharging cut-off voltage of 1.2 V. Only one layer separator system was used intentionally to make the cell have a internal short circuit. The charge efficiency of 93% - 99% was maintained until 18 cycles and started to decrease dramatically at 19 cycles owing to a internal short circuit. This sudden drop could be confirmed by the fluctuation and gradual declination of the Zn electrode potential with respect to Hg/HgO reference electrode at the end of charging period. During the cycle test, the impedance measurements were carried out at the specified cycle number for the charged and discharged Zn electrode (see Fig 8 and Fig 9) and the calculated results obtained by model I are listed in Table III and Table IV. It can be shown that the value of R3 for the charged electrode increases after the internal short circuit, on the other hand the value of C2 decreases. This can be explained that the amount of Zn metal decreases due to a lowering of charging efficiency. The value of R2 and R3 for the discharged electrode is always greater than that for the charged electrode because of the thickening of ZnO layer.

Fig 10 compares Zn electrode potentials as a function of SOC during charging period for several charge rates. During the initial charging period upto 4 0% SOC, Zn electrode potentials is more negative with increasing the charge rates, but the situation starts to be reversed at about 40% SOC and, moreover, reversd again during the end of charging The SOC at which the hydrogen evolution reaction proceed rigorously get higher in case of higher charging rates due to its lower charge efficiency.

Fig 11 shows the effects of charge current on the impedance spectra for the charged electrode in the condition of 100 % SOC and the calculated results are shown in Table V. The values of R_{ct}, C_{dl} and C_a were appeared not to have no correlation with increasing the charge rates, but that of R_a increased.

CONCLUSIONS

The impedance spectra measured for the electrode of low SOC were very different from that of high SOC, regardless of experimental conditions, such as charge rate, discharge rate and cycle number. Two types of equivalent circuit model were suggested to explain the impedance behavior at all experimental conditions. Model I which has a good agreement with the electrode of low SOC, contains three frequency domains; ZnO characteristics(HF), the corrosion reaction and double layer behavior(IF), and the diffusion process(LF). Model II for the electrode of high SOC, contains the corrosion reaction and double layer behavior (HF) and adsorption proces(LF). The impedance spectra is dramatically changed if the cell is short circuited. It is thus indicated that the impedance techniques can be a good diagonostics for evaluating the state of Ni-Zn battery.

REFERENCES

1. F.R. McLarnon and E.J. Cairns, J. Electrochem. Soc.,**138**, 645(1991)
2. J. Jindra, J. Power Sources, **37**, 297(1992)
3. G. Bronoel, A. Millot and N. Tassin, J. Power Sources, **34**, 243(1991)
4. K. Bass, P.J. Mitchell, G.D. Wilcox and J. Smith, J. Power Sources, **35**, 333(1991)
5. A.J. Bard, Electroanalytical Chemistry, Vol. 4, Marcel Dekker Inc., New York(1970)
6. M.A. Reid, J. Power Sources, **29**, 467(1990)
7. C. Cachet and R. Wiart, Electrochimica Acta, **29**, 145(1984)
8. J.P. Candy, P. Fouilloux, M. Keddam and H. Takenouti, Electrochimica Acta, **26**, 1029(1981)
9. N.A. Hampson and A.J.S. Mcneil, J. Power Sources, **15**, 61(1985)

10. J. Jindra and D. Becker-Rose, J. Power Sources, **43**, 55(1993)
11. J. Hendrikx, W. Visscher, E. Barendrecht, J. Appl. Electrochem., **16**, 175(1986)
12. J. Mcbreen and E. Gannon, Electrochimica Acta, **26**, 1439(1981)
13. L. Beaunier, I. Epelboin, J.C. Lestrade and H. Takenouti, Surface Technol., **4**, 237(1976)
14. A. Le Mehaute, *J. Stat. Phys.*, **36**, 665(1984)
15. D.W. Davidson and R.H. Cole, J. Chem. phys., **18**, 1417(1950)
16. R.D. Armstrong and M.F. Bell, J. Electroanal. Chem., **55**, 201(1974)
17. D.D. MacDonald, Transient Techniques in Electrochemistry, Plenum Press, New York, (1977)
18. J.L. Kuester and J.H. Mize, Optimization Technique with Fortran, McGraw-Hill Book Co., New York(1973)
19. F. Gutmann, J. Electrochem. Soc., **112**, 94(1965)
20. C. Lazarides and N.A. Hampson, Surf. Technol., **16**, 255(1982)
21. H. Gerischer, Z. Physik. Chem., **202**, 302(1953)
22. T.P. Dirkse and N.A. Hampson, Electrochimica Acta, **16**, 2049(1971)
23. N.A. Hampson and G.A. Herdman, J. Electroanal. Chem., **25**, 9(1970)
24. J. O'M. Bockris, Z. Nagy and A. Damjanovic, J. Electrochem. Soc., **119**, 285(1975)
25. Y.C. Chang and G. Prentice, J. Electrochem. Soc., **132**, 375(1985)
26. D.L. Cabot, M. Cortes, J. Electroanal. Chem. , **201**, 85(1986)

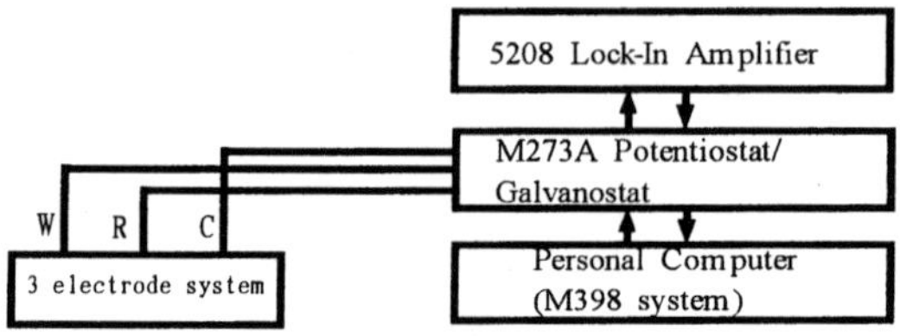

Fig. 1. Schematic diagram of experimental setup for impedance measurements.

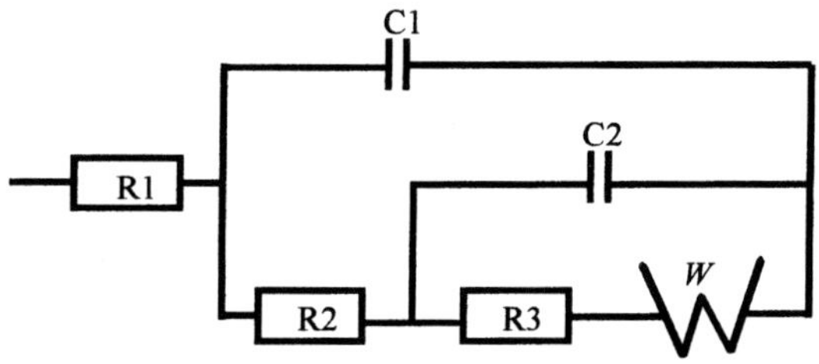

Fig. 2. Simple electrical equivalent circuit for Zn/ZnO layer/electrolyte system.

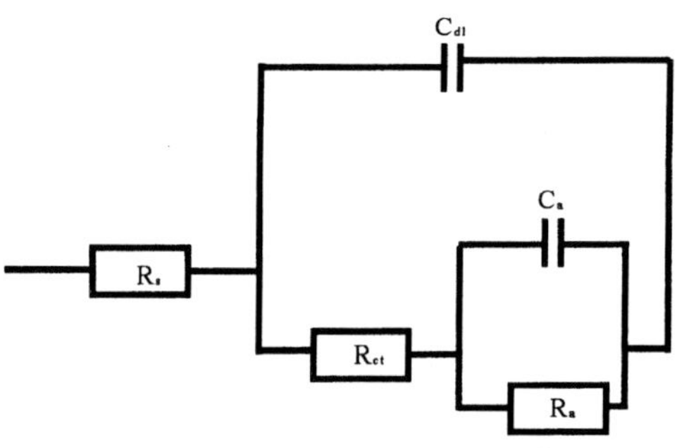

Fig. 3. Equivalent circuit for a adsorbed intermediate.

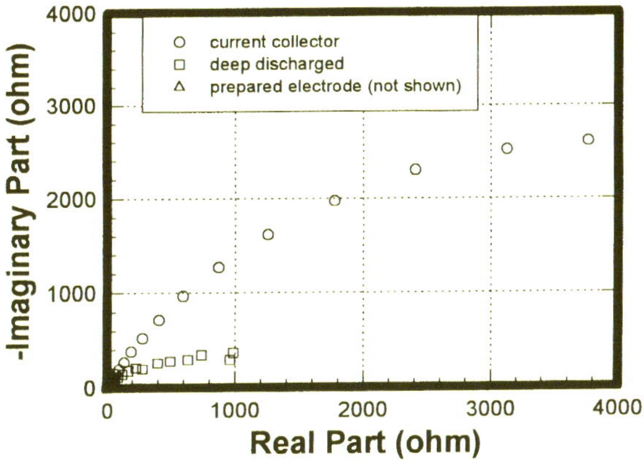

Fig. 4. Effect of deep discharge on impedance spectra.

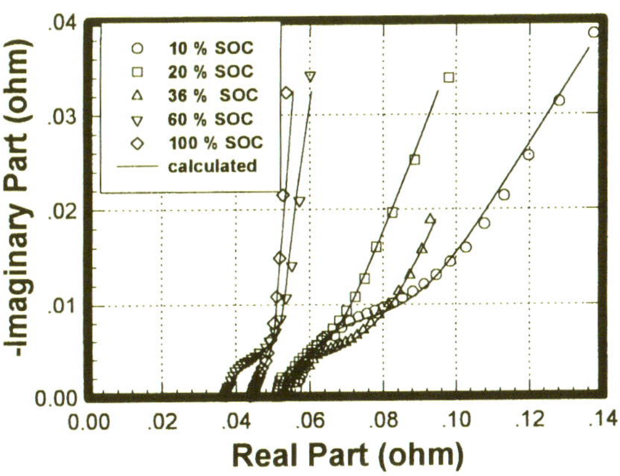

Fig. 5. Effect of SOC on impedance spectra.
(symbol: experimental, line: calculated)

Table I. Values of circuit elements obtained by modeling the impedance spectra of zinc electrode in various SOC

SOC	R1(ohm)	R2(ohm)	R3(ohm)	C1(F)	C2(F)	W (ohm $sec^{-1/2}$)	model
10 %	0.058	0.0035	0.084	16.8	1.08	0.019	I
20 %	0.051	0.005	0.01	1.97	33.1	0.006	I
36 %	0.053	0.013	0.023	5.8	77.4	0.004	I

SOC	R_s (ohm)	R_{ct} (ohm)	R_a (ohm)	C_{dl} (F)	C_a (F)	model
60 %	0.036	0.03	51000	65	9800	II
100 %	0.044	0.011	52000	236	9300	II

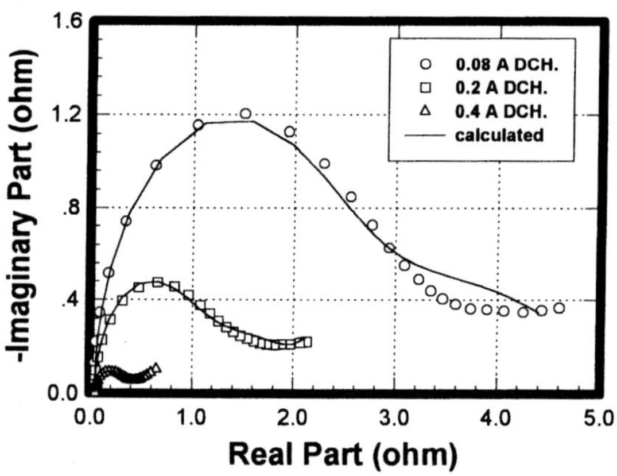

Fig. 6. Impedance spectra of zinc electrode discharged by various discharge rates. (symbol: experimental, line: calculated)

Table II. Values of circuit elements obtained by modeling the impedance spectra of zinc electrode discharged by various discharge rates

DCH. rate	R1(ohm)	R2(ohm)	R3(ohm)	C1(F)	C2 (F)	W (ohm $sec^{-1/2}$)	model
0.08 A	0.046	1.6	3.58	0.0103	0.35	0.00059	I
0.2 A	0.045	0.67	1.5	0.0101	0.53	0.027	I
0.4 A	0.049	0.00324	0.63	1.12	0.0183	0.021	I

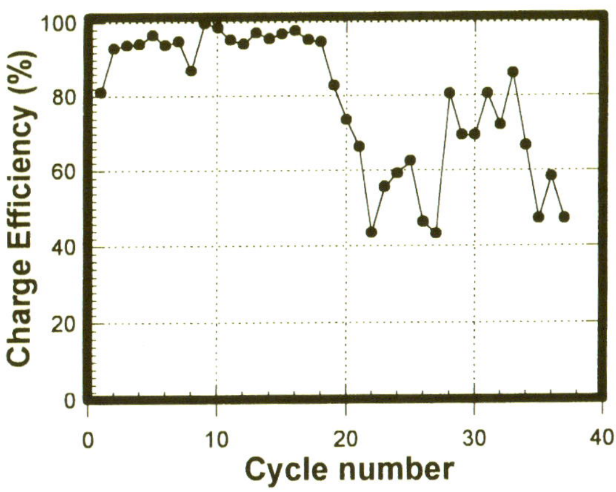

Fig. 7. Charge efficiency vs. cycle number.

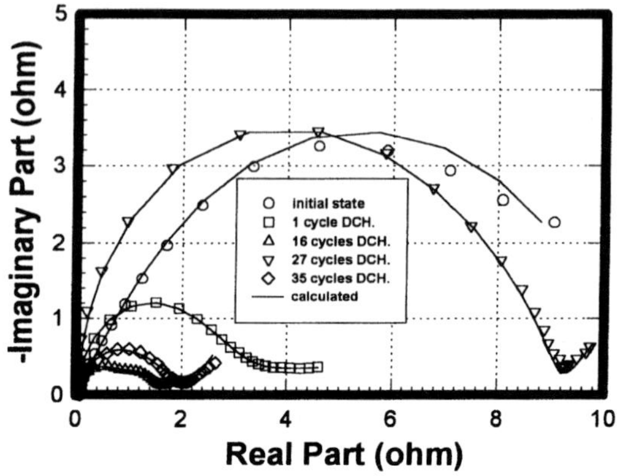

Fig 8. Effect of cycle number on impedance spectra of discharged zinc electrode. (symbol: experimental, line: calculated)

Table III. Values of circuit elements obtained by modeling the impedance spectra of discharged zinc electrode after specified cycle

cycle No.	R1(ohm)	R2(ohm)	R3(ohm)	C1(F)	C2(F)	W (ohm $sec^{-1/2}$)	model
0	0.04	1.38	8.6	0.12	0.176	6.7 X 10^{-7}	I
1	0.046	1.6	3.58	0.01	0.35	0.0006	I
16	0.044	0.588	1	0.002	0.14	0.040	I
27	0.055	6.23	2.8	0.0143	0.033	0.116	I
34	0.078	0.63	1.38	0.0122	0.063	0.092	I

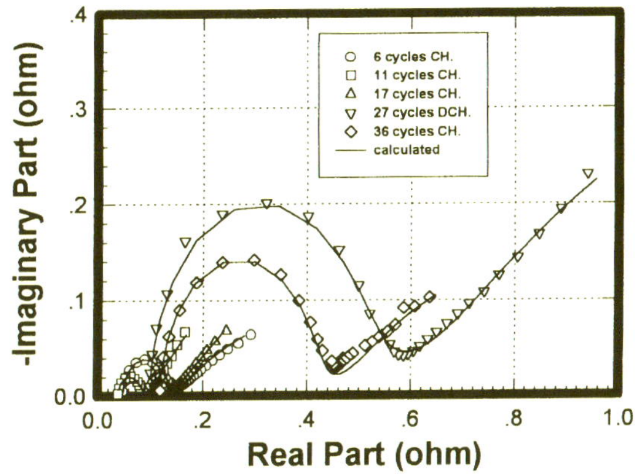

Fig. 9. Effect of cycle number on impedance spectra of charged zinc electrode (symbol: experimental, line: calculated).

Table IV. Values of circuit elements obtained by modeling the impedance spectra of charged zinc electrode after specified cycle

cycle No.	R1(ohm)	R2(ohm)	R3(ohm)	C1(F)	C2(F)	W(ohm sec$^{-1/2}$)	model
6	0.049	0.0225	0.196	0.003	5.25	0.074	I
11	0.039	0.0122	0.0531	0.0083	3.8	0.016	I
17	0.095	0.036	0.059	0.0085	24.7	0.074	I
27	0.149	0.074	1.04	0.793	0.0027	0.15	I
35	0.078	0.0023	0.77	1.51	0.0028	0.118	I

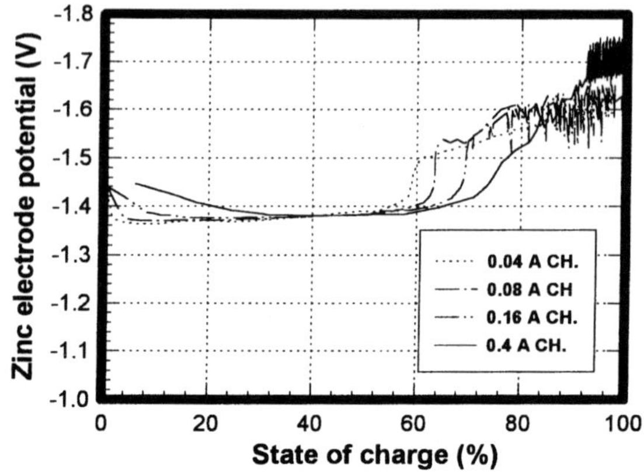

Fig. 10. Zinc electrode potential/SOC profile for various charge rates.

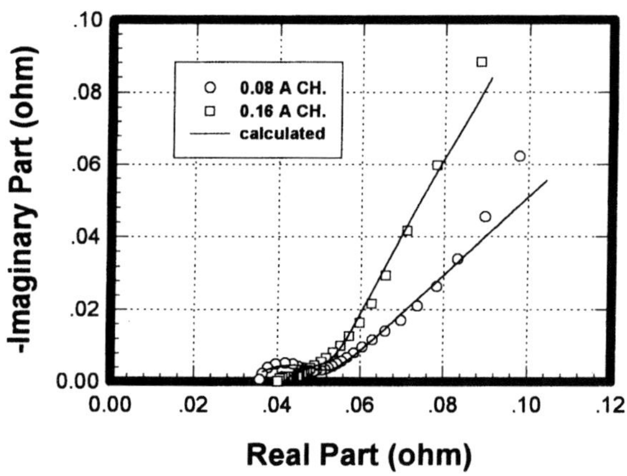

Fig. 11. Impedance spectra of 100% charged zinc electrode by various charge rates (symbol: experimental, line: calculated).

Table V. Values of circuit elements obtained by modeling the impedance spectra of 100% charged zinc electrode by various charge rates

CH. rate	R_s (ohm)	R_{ct} (ohm)	R_a (ohm)	C_{dl} (F)	C_a (F)	model
0.04 A	0.044	0.011	52000	236	932	II
0.08 A	0.033	0.02	64000	1.64	110	II
0.16 A	0.04	0.012	70700	21	125	II

PASSIVATION OF ZINC ELECTRODE IN THE SEALED NICKEL - ZINC CELLS

H. Chang, Y. B. Pyun, and M. E. Alexeeva*

Energy & Environment Laboratory
Materials & Devices Research Center
Samsung Advanced Institute of Technology
P. O. Box 111, Suwon, Korea

*Battery Company Rigel
St. Petersburg, Russia

ABSTRACT

Experiments have been performed on the nickel-zinc cells in order to determine the practical conditions for minimizing passivation of the Zn during the cycling. The influence of KOH, LiOH and zincate ion concentration on passivation of the Zn electrode have been investigated. Additives to the negative electrode were also studied. Highly alkaline electrolyte can prevent the Zn electrode from passivating due to its potential that is away from passivation. However, the passivation potential was lowered and the passivation current density increased so that the passivation started early, and a relatively thick passive film was formed in 8.5M KOH whereas a less thick film formed in 1M KOH. Addition of 4 wt.% Hg to the Zn electrode retarded passivation. The activation energy for passivation film growth has been determined and an electrolyte composition yielding slower kinetics has been suggested. A method of extending the life of the Zn electrode in the sealed nickel-zinc cells has been proposed by utilizing an optimum condition for preventing the passivation.

INTRODUCTION

Passivation of Zn in Alkaline Solution

Passivation is defined as protecting metal from corrosion by pumping electrons into a metal and thereby decreasing the potential difference across the metal - environment interface. The double layer field is altered to diminish the metal dissolution reaction that results in corrosion. The unstable surface of an actively corroding metal can be made passive and stable by increasing the potential in the positive direction and this reaction depends on the metal and solution composition.

Many investigations utilizing the galvanostatic method have been carried out and the Zn passivation mechanism, morphologies of passivation films and current densities for this reaction were reported (1). Various theories regarding the passivation of Zn in alkaline solutions have been suggested including the 'dissolution -precipitation model' proposing a reaction at potential below the critical value for the initiation of the solid state reaction (2,3), the 'adsorption model' and the 'nucleation and growth model'. It has also been concluded that no single mechanism explains the whole process (4,5). In all mechanisms the layer of passivation occurs at a critical potential ($E_{PASSIVATION}$, E_P) with its own current density ($i_{PASSIVATION}$, i_P). Most of these investigations on passivation under several conditions, however, have been carried out in the single component alkaline electrolyte (6~8).

Passivation of Zn in the Nickel-Zinc Cell

Together with its high specific energy, power density and the relatively low material cost, the nickel-zinc cell has its classical problem of short cycle life owing to cell shorting by Zn dendrite growth, and capacity loss by redistribution of Zn material causing shape change. One another reason for short cycle life that can not be ignored is passivation. Once the active Zn is passivated during discharge, especially on over discharge, and not recovered properly (no reports of activating the passivated electrode, yet), the amount of active sites on the Zn electrode is diminished. Thus, the charging efficiency is decreased so that the cycle life of the cell is limited. Currently, many methods for increasing cycle life is being developed, such as preventing the negative electrode from shorting by dendritic growth and capacity loss by shape change utilizing a new type of separator, low alkalinity electrolyte (9) and additives to electrode and current collector (10) and so on. Methods for suppressing the passivation of the Zn electrode in the cell have not been studied very much.

Passive films are formed on the Zn electrode under the specific conditions, i.e., in an alkaline solution in the anodic polarization region of the passive voltage range that is positive to -1.0 V (-1.2 V for the porous electrode (11)). The reaction is represented as follows by the dissolution-precipitation model which is one of the suggested mechanisms :

$$Zn + 4OH^- = Zn(OH)_4^{2-} + 2e \qquad [1]$$
$$Zn(OH)_4^{2-} = ZnO + H_2O + 2OH^- \qquad [2]$$

In a vented cell, passivation is more serious due to the fact that periodic maintenance of deep discharging down to the more positive potential over E_P is required. However, the sealed cell is free of maintenance, which does not bring about over discharge during the operation. Due to this fact, making a cell sealed itself is one of the ways of preventing passivation and this was achieved by utilizing additives in the active material for reducing hydrogen evolution with the aid of an appropriate charging method that can minimize the internal pressure build-up.

In the present half cell studies, electrolyte systems which are similar to practical sealed nickel-zinc cells with specific additives were studied. Some of the experiments have been performed in the actual battery system by the utilization of porous Zn electrode in order to find out the practical conditions for minimizing the passivation during cycling of the sealed cell. The influence of KOH, LiOH and zincate ion concentration of the electrolyte on the passivation of the Zn electrodes have been investigated in the half cell and sealed nickel-zinc test cells. Each cell has capacity of 3.5 Ah with the negative and positive electrode capacity ratio C(-)/C(+) = 3. The rate of growth of the passivation film is studied. Additions such as HgO, PbO, CdO and In_2O_3 to the negative electrode were also studied.

EXPERIMENTAL

The cyclic voltammogram and anodic polarization curves were obtained by a Potentio-Galvanostat using Zn as the working electrodes, a Hg/HgO electrode as the reference, and Ni as the counter electrode. All the electrode potentials in this half cell experiment were referred to Hg/HgO electrode. The working electrode, Zn was obtained as 99.99 % pure rods (Aldrich, USA) and was into a pellet and mounted in epoxide resin. The exposed area of electrode was 1.327 cm^2 after repeated polishing. Prior to each run, the working electrodes were ground with 600 grit silicon carbide followed by further polishing with 0.05 μm alumina powders. The influence of KOH (1, 5.5, 7 and 8.5M), LiOH (0.6, 1 and 1.25M) and zincate ion concentration (0.1, 0.25 and 0.5M) in the electrolyte was investigated. Reagent grade KOH (Junsei, Japan), $LiOH\text{-}H_2O$ (Aldrich, USA) and ZnO (Hanil, Korea) with distilled deionized water were used for preparing these electrolytes.

Electrochemical data were obtained using a Potentio-Galvanostat (EG&G Princeton Applied Research, M273). The cyclic voltammetry studies were made in voltage envelopes of -0.4 V to -1.8 V at the sweep rate of 5 mV/sec. Anodic polarization curves were obtained by the potentiodynamic method, at a sweep rate of 0.2 mV/sec. The voltage scanning experiments were performed at 30, 40, 50 and 60℃.

Polymer bonded ZnO powder was roll pressed onto a Cd coated Cu mesh in the dry state to obtain a negative electrode. Appropriate amounts of the additives were mixed to the electrode during mixing the active mass or amalgamated in the wet state. Positive electrodes were prepared by impregnation of nickel hydroxide into sintered nickel plaques. Sets of electrodes were assembled with several layers of organic microporous film and filled with electrolyte in a test cell. These cells (10) having 3.5 Ah capacity were charged at the C/12 rate and discharged at the C/3 rate on two groups of test cells down to the limit voltage of 1.2 V and part of them were over discharged to 0.8 V in order to accelerate the passivation. Here the potential was referred to the actual cell voltage. During cycling the test cells, discharge capacity was monitored. After the capacity of the test cells decreased down to 60 % of the initial capacity, the test cells were disassembled in the discharged state and the phase composition (Metal Zn/ZnO) of the negative electrode active masses were determined. The influence of additives (HgO, PbO, CdO and In_2O_3) into Zn electrode on passivation was also studied.

RESULTS AND DISCUSSION

Cyclic Voltammetry

Figure 1(a) shows a typical cyclic voltammogram of the oxidation (discharge) and reduction (charge) of Zn in an alkaline solution, which summarizes a passivation (point A) and breaking the oxidation layer (point B) during the cell operation. Figures 1(b), (c) and (d) show the passivation characteristics at the various elelectrolyte conditions. As the potential was made more positive starting from corrosion potential, the dissolution current increases. However, at certain point, the current - potential curve changes direction and the current decreases rapidly with potential. The KOH concentration affected the E_P severely so that it came down to -1.045 V at an electrolyte concentration of 8.5M KOH + 1M LiOH + 0.5M ZnO. In the 1M KOH electrolyte the passivation started late since the E_P was much more positive, -0.943 V and was not affected much by the concentration of LiOH and zincate ion. The growth rate of passive film could be obtained from the i_P and calculated detail from the anodic polarization curves.

Anodic Polarization Curve

The curves in Figure 2 show the anodic polarization of the Zn electrode in 7M KOH + 1M LiOH + 0.5M ZnO at elevated temperatures. Approaching the passivation potential, the dissolution process was inhibited. At this stage of the transition region, a prepassive, unstable corroding film was formed, which occurred at a potential still negative to the E_P and is ionically conductive. As the potential moves positively to the passive region, the dissolution stopped so that Zn became passive and the surface was stabilized where the passivation potential and current is defined. For the case of other electrolyte systems at various KOH, LiOH and zincate concentration, the same trends were attained.

Shown in Figure 3 is the effect of KOH concentration on the E_P. It can be seen that the characteristics at room temperautre are different from those at high temperatures. The E_P decreased as the KOH concentration increased proving that the highly alkaline electrolyte might motivate passivation more easily. However, when the concentration is higher than 5.5M KOH above 40℃ or above, the E_P increased by increasing alkalinity. An explanation has not been suggested here, but it is thought that alkalinity at high temperature causes dissolving the passivated ZnO films.

Passivation Kinetics

By assuming that the passivating layer consists of ZnO, the growth rate of the layer was calculated from the data of anodic polarization. If the current densty is given from the polarization curve, the weight gain by the deposition per time in the unit area can be calculated and the rate of Zn deposition is determined. Faraday's law may be summarized by

$$W = itM/zF \quad [3]$$

$$\delta = W/\rho = itM/\rho zF \quad [4]$$

$$R = \delta/t = iM/\rho zF = 10.5 \times i \ [\text{Å/sec}] \quad [5]$$

where, W the weight of the deposition in g/cm^2, i the passivation current density in A/cm^2, t the elapsed time in sec, M the molecular weight of the oxidized substance in g/mole, z the number of charges involved, F the Faraday's constant, δ the thickness of passivating layer and ρ the density of the passivating layer (5.7g/cm^3 when porosity is taken as 0.27) (12). And, R is the growth rate of passivating layer in Å/sec and the constant 10.5 is calculated for the case of ZnO passivation film. From this relationship, the growth rates of passivation film have been calculated for the case of various electrolyte conditions at elevated temperatures. Activation energies for the growth of passivation film at each electrolyte were attained from the Arrhenius plots and shown in Figure 4, which suggest that the film grows fast in the highly alkaline electrolyte since it requres lower acivation energy (17.33 kJmole^{-1}K^{-1} at 8.5M KOH and 24.34 kJmole^{-1}K^{-1} at 1M KOH).

The passivation current affected the passivating layer thickness and the effect of KOH, LiOH and zincate ion concentration was severe, thus an optimum concentration has to be determined for having an electrolyte that has both characteristics of high ionic conductivity and minimizing the passivation. Shown in Figure 5, the KOH concentration was suggested to be limited to 5.5M for sustaining low current density and high enough ionic conductivity. Even though the higher concentration of KOH could lower the cell voltage negative to the E_P because of the high activity so that it can prevent Zn electrode from passivation, the E_P itself decreased more steeply as KOH concentration increases (Figure 3).

The i_P data (Figure 5) were supplemented with microscopic examination. In the region near 8.5M KOH thick (> 500 Å) passive film was formed, whereas a less thick (< 100 Å) film formed in the case of 1M KOH. Due to this fact, an optimum concentration of KOH was determined in the matter of passivation as 5.5 M. As can be seen in the Figure 5, zincate ion dissolved in the electrolyte for inhibiting the shape change in the nickel-zinc cell also affected on the i_P. It was found that an elelctroytle with more than 0.25M of zincate ion led to a steep increase of i_P resulting in thickening the passivation film. Looking into the limiting current density, the effect of zincate ion was more severe. When the ZnO amount in the electrolyte increased to 80 g/l that is fully saturated, the limiting current decreased steeply down to ~15 mA/cm^2, so that the passivation could occur more actively comparing with 30 mA/cm^2 in the case of 20 g/l (0.25M). It was also found that there was not much effect by the LiOH concentration, but the i_P falled down as it was over 1M LiOH.

Additives in the Electrode

As to the passivation, it is a reaction between the electrode and its environment, so that the electrode condition is important factor to decide the reaction variables. With regard to sealing the nickel-zinc cell, controlling the internal gas pressure is essential and utilizing Hg in the negative electrode is widely known. This is also beneficial not only for yielding less hydrogen evolution, but for retarding passivation. A lower electrode potential was attained, which means more stable from passivation as the Hg content increased. Near 4 wt.% of Hg derived lowest cell potential (-1.48 V) and was good for preventing passive layer formation and also made it easy to break down the passive film. In the anodic polarization curve, the Hg affected the passive dissolution region with much of effect being observed in the passivation region (> -1.0 V). These results obey that the prepassive film was derived from adsorbed intermediate species when the electrode contains Hg. This had been confirmed by result that the rate determining steps in the active dissolution of Zn with and without Hg were different from each other due to an effect of ionic strength in the case of Hg added Zn electrode (13). More tests were performed with the additives such as PbO, CdO and In_2O_3 which can replace Hg. Not much of effect as from Hg was not obtained. However, addition of CdO and PbO showed compatible effects showing that the E_P is more positive (-1.0 V) than a cell with Hg in the Zn electrode (E_P = -0.95 V) (14). Passivation potential of the cell without any additives in the Zn electrode was -1.2 V. More studies with In_2O_3 are needed, but the possibility of removing Hg out of the Zn electrode for retarding the passivation was shown.

Conditions Providing Improved Cycle Life

An optimum condition for retarding the passivation for the sealed nickel-zinc cell has been determined. An electrolyte system with 5.5M KOH with 1M LiOH and 0.25M dissolved zincate ion was choosed for having optimized performance in the cell. Another electrolyte system which had been utilized for vented cell (8.5M KOH, 1M LiOH, 0.5M Zincate) was also prepared and compared with the optimized electrolyte system. A cell filled with the optimized electrolyte lasted longer cycle life close to 300 cycles and showed relatively higher Zn electrode capacity as the cell was fully discharged at the periodical measurements during the cycling. Original capacity of Zn electrode was 10.5 Ah in the cell with nominal capacity of 3.5 Ah. Addition of Hg in the electrode aided the cell to be completely sealed by reducing the H_2 gas evolution and helped to retard passivation of Zn electrode. Substitution of Hg by Pb in the Zn electrode was also successful either for reducing H_2 evolution and passivation control.

CONCLUSIONS

In the sealed nickel-zinc cell, highly alkaline electrolyte could prevent Zn electrode from passivation due to its cell potential that is away from passivation potential. However, cyclic voltammograms showed that the electrolyte with high KOH concentration lowered passivation potential more negatively and made the passivation start earlier. LiOH and zincate ion concentration did not change the passivation potential as much as KOH.

With regards to passivation current density which is related to the passivation film thickness, relatively thick film was created in 8.5M KOH that required a high value of passivation current density, whereas less thick film formed in 1M KOH. Anodic polarization curves were obtained at elevated temperatures and activation energy for the passivation film growth at each case of electrolyte were calculated. The lower activation energy in the electrolyte with higher KOH and zincate ion concentration proved its fast kinetics, which also led thicker passivation film.

Additives such as Hg, Pb and Cd in the Zn electrode were beneficial for retarding the passivation and it was also found that lower temperature reduced rate of thickening of the film.

Conditions for higher passivation potential, low passivation current density and slow kinetics of passivation film growth were investigated. An electrolyte with 5.5M KOH with 1M LiOH and 0.25M of dissolved zincate ion was choosed for having optimized performance with regards to the passivation control and maintaining the higher capacity. Thereby, a way of improving the cycle life of the Zn electrode in the sealed nickel-zinc cell has been proposed.

ACKNOWLEDGMENTS

One of the authors (H.C) thanks to Prof. Elton J. Cairns at the University of California at Berkeley (Lawrence Berkeley Laboratory) for his valuable comments.

REFERENCES

1. R. D. Armstrong and M. F. Bell, Electrochemistry Vol. 4, H. R. Thirsk, Editor, Ch.1. Chemical Society, London (1974).
2. R. W. Powers and M. W. Breiter, J. Electrochem. Soc., **116**, 685 (1969).
3. A. Marshall and N. A. Hampson, J. Appl. Electrochem., **7**, 271 (1977).
4. R. J. Brodd and V. E. Leger, Encyclopedia of the Electrochemistry of the Elements Vol.4, Ch.V-1, (1975).
5 F. L. Tye, Electrochemical Power Sources, M. Barak, Editor, p.128, Peter Pereginus Ltd., London (1980).
6. R. Armstrong, Corr. Sci., **11**, 693 (1971).
7. M.-B. Liu, G. M. Cook and N. P. Yao, J. Electrochem. Soc., **128**, 1663 (1981).
8. L. Baugh and A. Higginson, Electrochimica Acta, **30**, 1163 (1985).
9. T. C. Adler, F. R. McLarnon and E. J. Cairns, J. Electrochem. Soc., **140**, 289 (1993).
10. H. Chang and C. Lim, Power Sources, **15**, 139 (1995).
11. Private discussion with L. Reikhelson at Rigel, St.Petersburg (1994).
12. B. Aurian-Blajeni and M. Tomkiewicz, J. Electrochem. Soc., **132**, 869 (1985).
13. J. Hendrix, A. van der Putten, W. Visscher and E. Barendrecht, Electrochimica. Acta, **29**, 81 (1984).
14. Y. B. Pyun, H. Chang and M. E. Alexeeva, in Rechargeable Zinc Batteries Symposium, The 188th Electrochem. Soc. Meeting, Chicago, (1995) in print.

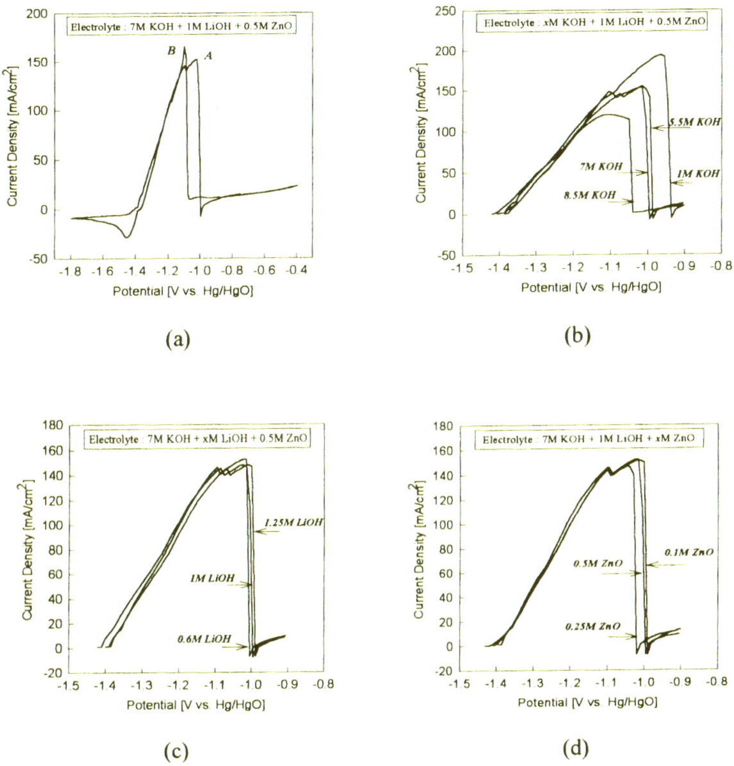

Figure 1. Cyclic voltammograms of Zn showing (a) a typical curve of oxidation and reduction of Zn in the alkaline electrolyte, (b) the effect of KOH concentration (c) the effect of LiOH and (d) the effect of zincate ion concentration on the passivation behavior

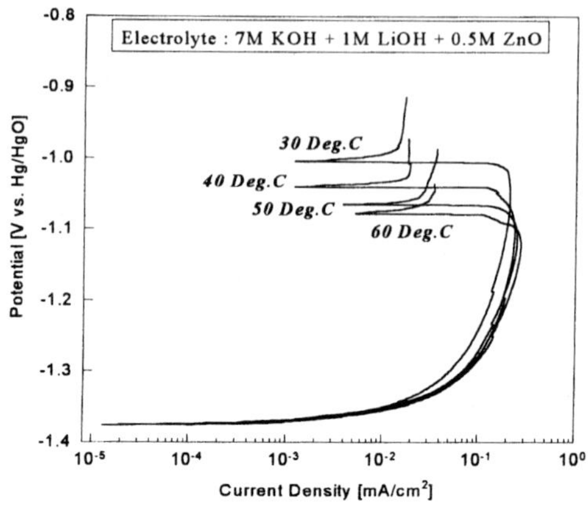

Figure 2. Anodic polarization curves of Zn in the alkaline electrolyte at elevated temperatures

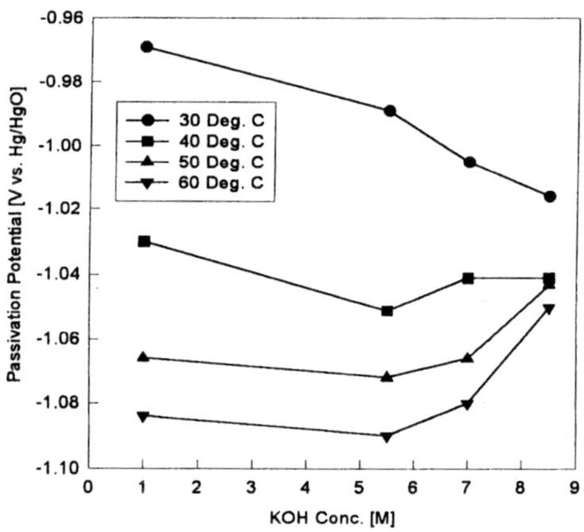

Figure 3. Passivation potential of Zn in various concentration of KOH electrolyte at the elevated temperatures

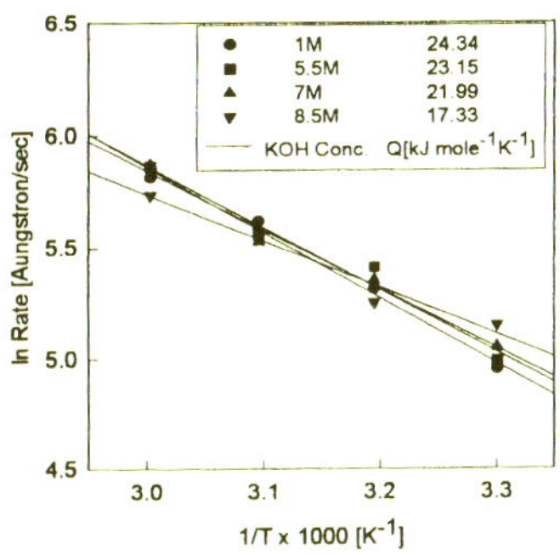

(a)

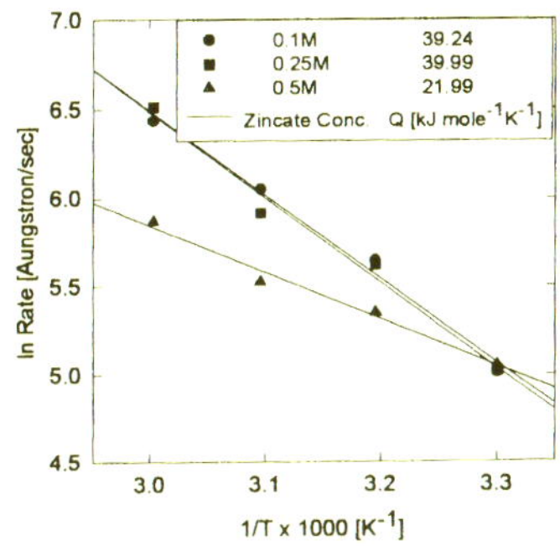

(b)

Figure 4. Arrhenius plots showing (a) KOH concentration effect (b) Zincate ion concentration effect

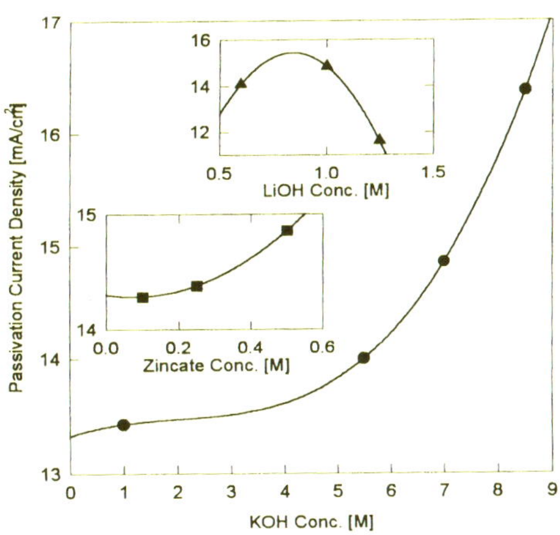

Figure 5. Passivation current density as a function of additives in the electrolyte

NOVEL ELECTROLYTE ADDITIVES TO ENHANCE ZINC ELECTRODE CYCLE LIFE

Narayan Doddapaneni and David Ingersoll
Sandia National Laboratories
Battery Research Department
MS-0614
Albuquerque, NM 87185-0614, USA

Introduction

Electrochemical power sources that utilize zinc electrodes possess many advantages. Zinc is abundantly available, benign, inexpensive, stable over a wide operating temperature range, and has a high oxidation potential. In spite of these advantageous characteristics, rechargeable electrochemical systems based on zinc chemistry have not found widespread use. The major disadvantages of zinc electrodes are that power sources with zinc electrode have limited cycle life due to zinc slumping and zinc electrode shape changes in alkaline solutions resulting from the solubility of zincate ($Zn(OH)_4^{2-}$) in these solutions. As a result, premature cell failure often results due to cell shorting caused by dendritic growth as well as zinc slumping [1,2].

In this paper we describe the chemical and physical characteristics of electrolyte solutions employing additives, particularly for zinc based electrochemical systems. These electrolytes are prepared using the alkali metal salts of 1,3,5-phenyltrisulfonic acid in combination with potassium hydroxide [3]. The alkali metal salts of the acid possess good thermal stability, good ionic conductivity, and have a wide electrochemical voltage window in aqueous and non-aqueous systems. With these electrolyte solutions improved cycle life was achieved in Zn/NiOOH and Zn/AgO. Improved cycle life with this additive is attributed to decreased zincate solubility, resulting in reduced zinc slumping and electrode shape changes. In addition, increased shelf-life and reduced self-discharge were also observed in many alkaline power sources.

Experimental

Additives, such as sodium and potassium salts of 1,3,5-phenyltrisulphonic acid were prepared in our laboratory using the standard procedure. All solutions were prepared using deionized water. Potassium hydroxide solutions of the desired concentrations were first prepared, and then various amounts of the potassium salt of potassium 1,3,5-phenyltrisulphonate (KPTS) were added to obtain the final desired concentration of salt in 15% and 20% KOH solutions. Saturated solutions of potassium 1,3,5-

phenyltrisulphonate in 42% KOH solutions were prepared by vigorously stirring excess potassium 1,3,5-phenyltrisulphonate in 42% KOH at room temperature for 24 hours. This resulted in less than 1% potassium 1,3,5-phenyltrisulphonate in solution. It was found that the solubility of potassium 1,3,5-phenyltrisulphonate decreases with increased concentrations of KOH. Conductivity measurements of solutions were made using a standard liquid conductivity cell having a cell constant of one. Thermal gravimetric analysis were performed using a Perkin Elmer TGA 7 at scan rates of 2°C/min. Electrochemical measurements were performed using a BAS 100A potentiostat and freshly polished glassy carbon electrodes. All measurements were made in deoxygenated solutions. Zn/NiOOH and Zn/AgO cells were built by Westinghouse and Yardney Technical Products, respectively. The solutions used in the Zn/NiOOH cells was composed of either 15% or 20% KOH containing 29% potassium 1,3,5-phenyltrisulphonate, while the Zn/AgO cells were filled with a solution of 42% KOH saturated with potassium 1,3,5-phenyltrisulphonate. Celgard 3401 and lead grids for the zinc electrodes were used in both systems.

RESULTS AND DISCUSSION

Physical Properties

In Figure 1, the conductivity of potassium 1,3,5-phenyltrisulphonate as a function of concentration is shown. For example, the conductivity of a 1M solution of potassium 1,3,5-phenyltrisulphonate is 0.095 S/cm, as compared to 0.17 S/cm for 1M KOH. Thermogravimetric analysis of sodium 1,3,5-phenyltrisulphonate (NaKPTS) in the presence of oxygen as well as in inert atmospheres was conducted. In Figure 2 is shown the TGA of NaKPTS in an oxygen atmosphere. Finally, its electrochemical behavior was evaluated at a freshly polished glassy carbon electrode, and the linear sweep voltammograms recorded at 50 mV/min is shown in Figure 3. The working range of electrolyte solutions prepared using this salt are as good as many other known salts, and is limited by the oxidation and reduction of water.

Electrochemical Cell Performance

The effect of the presence of potassium 1,3,5-phenyltrisulphonate additive in KOH solutions were evaluated in both Zn/NiOOH and Zn/AgO test cells. The Zn/NiOOH cells (8Ah) were flooded cells and were zinc limited. These cells were filled with electrolytes of either of two compositions: 1) 29% potassium 1,3,5-phenyltrisulphonate in 15% KOH or 2) 29% potassium 1,3,5-phenyltrisulphonate in 20% KOH. These electrolytes have conductivities comparable to the 30% KOH electrolyte (0.61 S/cm) which is commonly used in consumer batteries. The cells were charged at 12.5 mA/cm^2 to 1.89V and discharged at 25 mA/cm^2 to 1.35V. The performance of these cells is shown in Figure 4. After 50 cycles the cell performance started degrading and after 70 cycles the cells were removed from test and physically examined. It was found that the lead wires, which were attached using Pb/Sn solder, had corroded. The leads were reattached using molten Pb. After replacing the separators with new separators the cells were reassembled using the same electrodes, and were then placed back on test. Small amounts of KOH solutions were added in order to wet the new separators. As can be seen in the figure, the cells returned to their original performance after reassembling. After 106 cycles, with minimal loss of capacity, these cells were removed from test and postmortem evaluation performed. A small amount of zinc was found on the bottom of the cell, with no observable shape changes. The electrolyte solutions were analyzed and no degradation products were found.

The additive was also found to help in reducing self-discharge rates. The open circuit loss of cells rated at 8Ah after 120 hours at 24°C is given in Table 1. Average losses of 12% over five days was found. These losses are comparable to nickel-iron, but less than one-half that typically found for nickel-cadmium, nickel-metal hydride, or nickel-zinc cells with 30% KOH solutions. It was also found that addition of small quantities of potassium phenyltrisulphonate to alkaline solutions generally reduces the self discharge in most of the alkaline power sources.

Table 1. Open circuit loss of Zn/NiOOH cells at 24°C.

cell number	capacity - previous cycle (Ah)	capacity after 120 hrs (Ah)	% loss
1	5	4.4	12%
2	5.1	4.4	14%
3	3.7	3.3	11%
4	3.5	3.1	12%

Three Zn/AgO cells, rated at 12 Ah, were filled with 42% KOH and served as controls. Three Zn/AgO cells were filled with 42% KOH saturated with potassium 1,3,5-phenyltrisulphonate. All of the cells were charged at 0.5A to 2.05 V and discharged at 4 A to 1.1 V. The average performance of these cells is depicted in Figure 5. The cell-to-cell performance variation is found to be very small. As can be seen, the performance of all of the cells are about the same up to 70 cycles. However, at higher cycle numbers the capacity fade rate appears to be smaller for those cells containing the additive. At the present time over 80 cycles have been achieved, and they are continuing to be cycled.

Summary and Conclusions

Novel electrolyte additives were evaluated in Zn/NiOOH and Zn/AgO cells. Initial results indicate enhanced cycle life of these cells. The additive appears to reduce the zinc oxide solubility and thereby increases the zinc electrode cycle life. For example, the solubility of ZnO in 15% KOH is 0.35 M [4], whereas the solubility of ZnO in 15% KOH containing 29% potassium 1,3,5-phenyltrisulphonate is only 0.06 M. The additive also seems to be adsorbed onto the surface of the zinc electrode preventing and/ or inhibiting zinc electrode shape changes and thereby enhancing the stability of the zinc electrode structure. The self discharge rate of Zn/NiOOH cells are dramatically reduced in the presence of potassium 1,3,5-phenyltrisulphonate additive. Further experiments are in progress to understand the nature of the interfacial coating and zinc electrode morphology changes with cycle life. The effect of the additive on the cyclic voltammetric behavior at zinc electrodes is being studied.

Acknowledgments

The authors wish to express their thanks to Dr. J. Jackovitz of Westinghouse Electric Corp and Mr. R. Serenyi of Yardney Technical Products Inc. for their contributions to this paper.

References

1. F.R. McLarnon and E.J. Cairns, *J. Electrochem. Soc.*, **138** , 645, (1991).
2. M. Klein and F.R. McLarnon, "Nickel Zinc Batteries", in Handbook of Batteries, 2nd. Ed., D. Linden editor, McGraw Hill Inc., p 29.1, (1995).
3. N. Doddapaneni and D. Ingersoll, US Patent 5,378,550 (Jan. 3, 1995).
4. T.P. Dirkse, *J. Electrochem. Soc.*, **115**, 784, (1968).

Figure 1. Conductivity of Solutions of Potassium 1,3,5-phenyltrisulphonate as a Function of Molarity

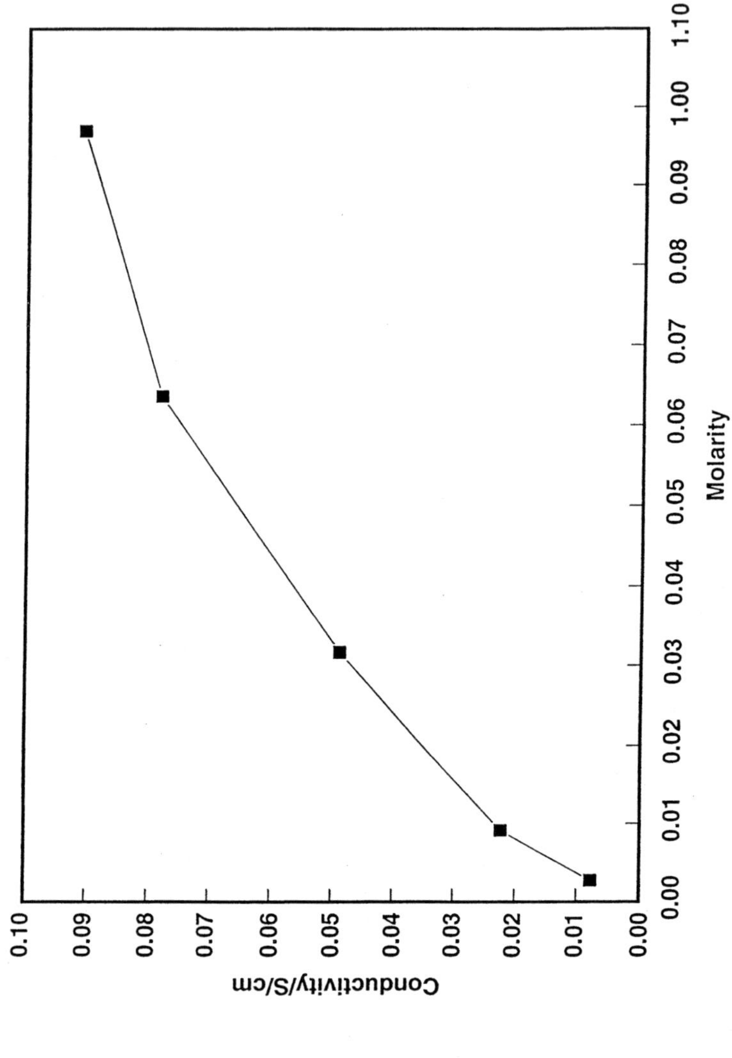

2629.01

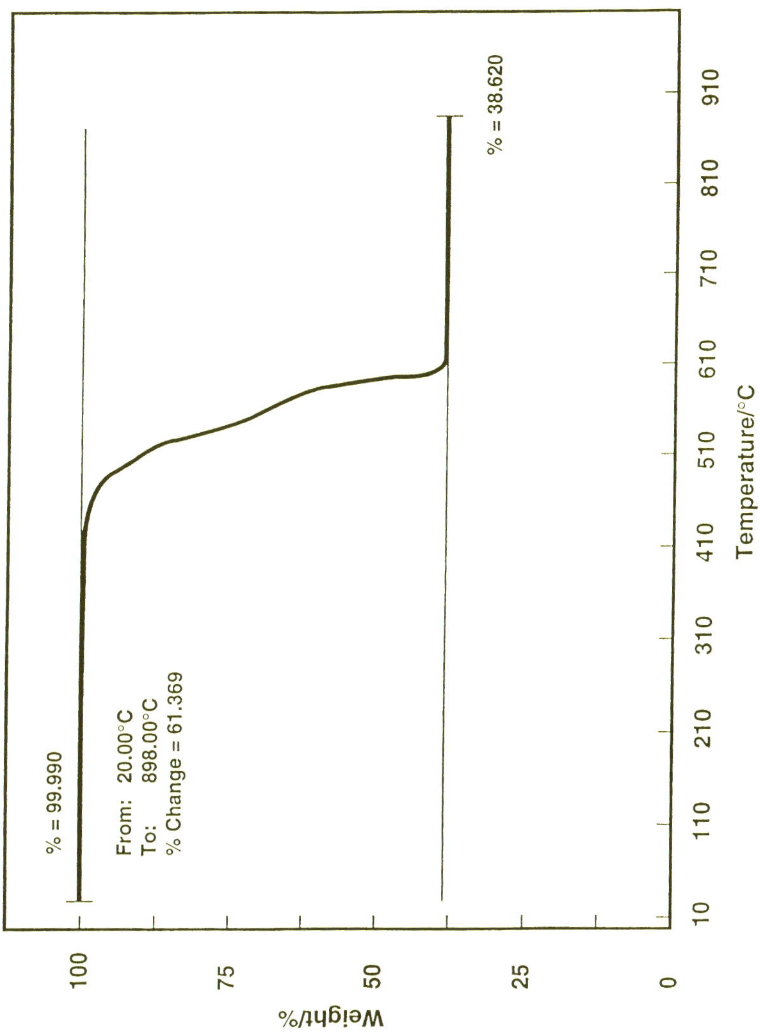

Figure 2. Thermal Gravimetric Analysis of Sodium 1,3,5-phenyltrisulphonate in Oxygen at 2°C/min

5629.02

Figure 3. Linear Sweep Voltammograms of Sodium 1,3,5-phenyltrisulphonate

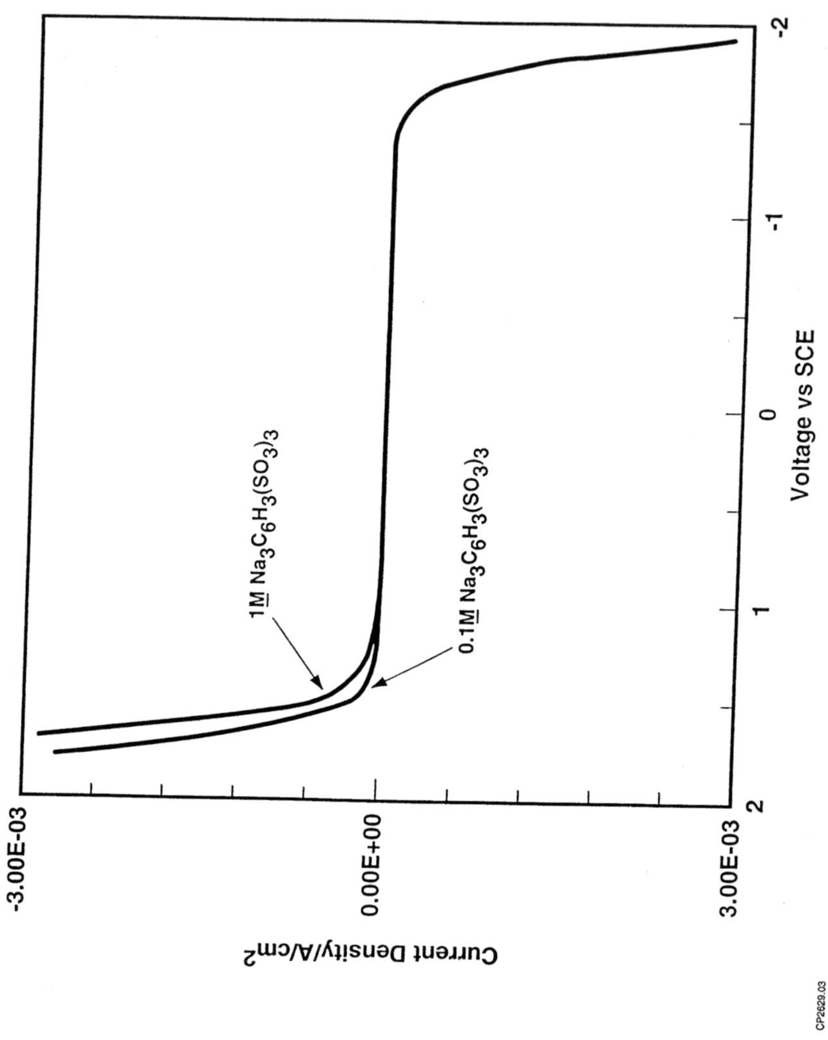

Figure 4. Performance of Zn/NiOOH Cells With Westinghouse Paste Zinc Electrodes in KOH Solutions Containing Potassium 1,3,5-phenyltrisulphonate Additive

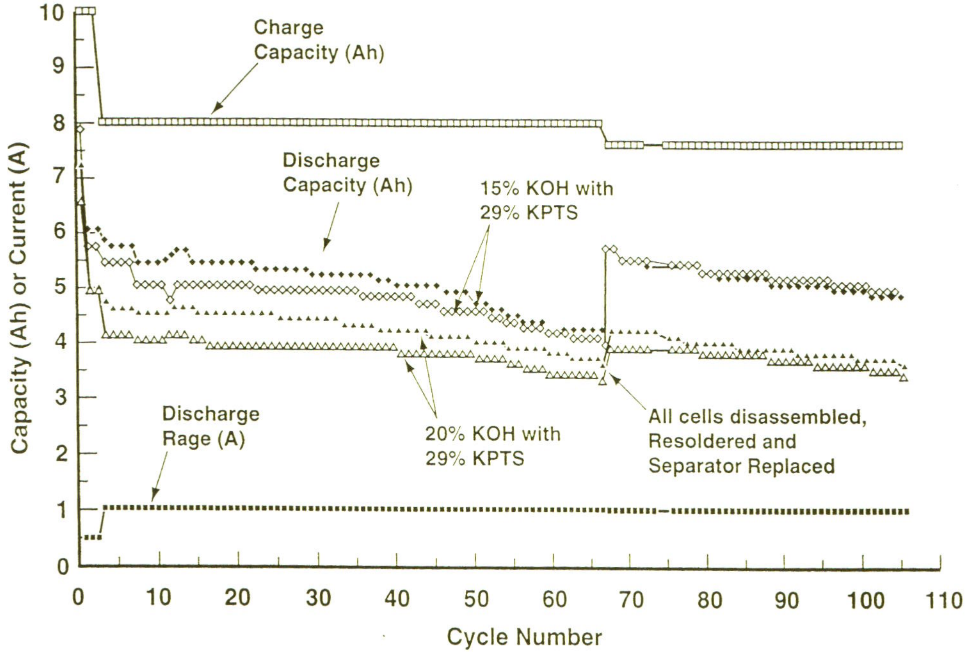

Figure 5. Performance of Zn/AgO Cells

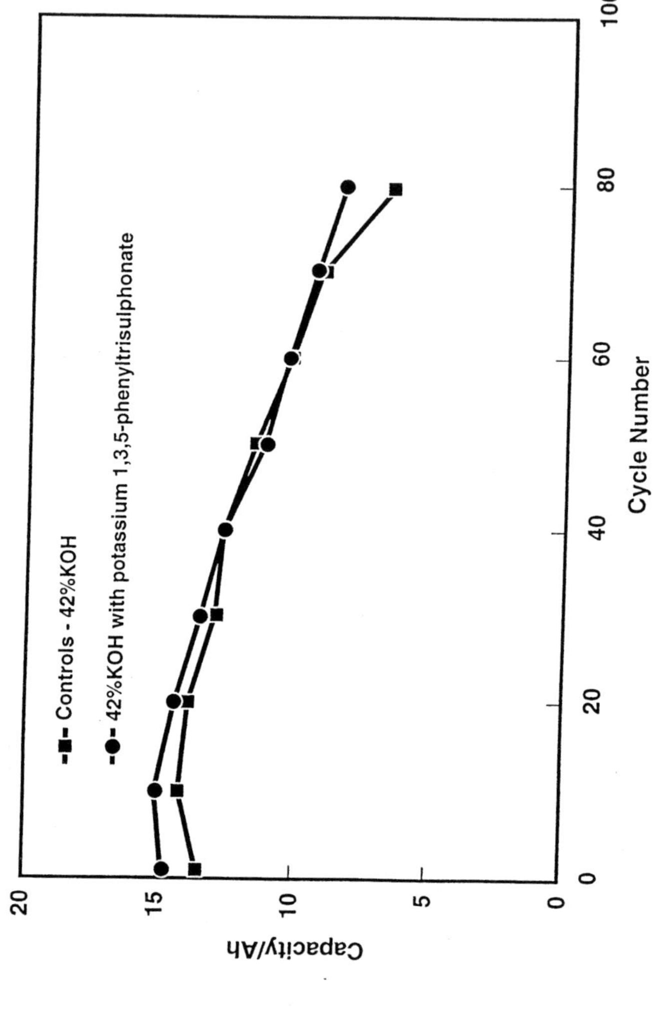

629.05

OXYGEN IONIZATION IN THE SEALED NICKEL-ZINC CELLS

Y. B. Pyun, H. Chang and M. E. Alexeeva*

Energy & Environment Laboratory
Materials & Devices Research Center
Samsung Advanced Institute of Technology
P.O. Box 111, Suwon, Korea

* Battery Company Rigel
St. Petersburg, Russia

ABSTRACT

Various factors which affect the O_2 gas ionization in sealed Ni-Zn cells have been investigated to determine the conditions for improving the O_2 gas ionization. Increasing the initial State-of-Charge of Zn electrode, i.e., the amount of metallic Zn, improves the rate of O_2 gas ionization. Addition of HgO, PbO and In_2O_3 is effective for higher rate of the O_2 gas ionization so that the cells maintains the low O_2 gas pressure during the cycling. This is due to suppressing the H_2 gas evolution from Zn electrode and retarding the passivation of metallic Zn. When surface of metallic Zn is prevented from the passivation throughout the cycling, the O_2 gas ionization occurs actively on the Zn electrodes.

INTRODUCTION

The secondary Ni-Zn batteries have long been considered as a power source for electric vehicles (1). Even with improving its short cycle property, there are, still, the problems related with the internal gas pressure and periodic maintenance. In order to solve these problems the internal gas pressure build-up by the evolution of O_2 gas and H_2 gas from electrodes during the operation of the Ni-Zn cells should be avoided.

The H_2 gas evolves at the Zn electrode when the Zn electrode is overcharged. However, the H_2 gas evolution from overcharged Zn electrode would be prevented if the capcity of Zn electrode is much higher than that of Ni electrode so that the Ni electrode is fully charged before the Zn electrode reaches full charge (2). Another reason causing the H_2 gas evolution from Zn electrode is the corrosion of metallic Zn in alkaline solution (2-5). Many corrosion inhibitors for suppressing the H_2 gas evolution have been proposed and showed the desirable effect of the lowering H_2 gas pressure for sealed Ni-Zn cells (6-11).

Since the O_2 gas evolution at Ni electrode causes the increase of cell pressure during the charging, it is necessary to minimize the O_2 gas evolution and remove the evolved O_2 gas in the cells. The principles for achieving sealed operation in the sealed Ni-Cd cells have been investigated (12-14). As to sealed Ni-Cd cells, fast O_2 gas diffusion from Ni elcetrode to Cd electrode is much easier due to the separator structure of the Ni-Cd cells. Thus, it is possible to maitain the low cell pressure by reducing the evolved O_2 gas at the surface of metallic Cd. This O_2 gas ionization in the Cd electrode is represented as following (15) :

$$O_2 + 2H_2O + 4\,e^- \rightarrow 4OH^- \qquad [1]$$

Even though metallic Zn is more active electrochemically than metallic Cd, the O_2 gas ionzation is actually difficult to occur in sealed Ni-Zn cells because that the gas-nonpermiable film separator should be used for preventing the dendritic Zn growth to Ni electrode (2,16). In addition, because the O_2 gas ionization in the sealed Ni-Zn cells is affected not only by the O_2 gas diffusion behavior but by elctrochemical properties of metallic Zn, the O_2 gas ionization on Zn electrode is more complicated and its rate is slower than that on the Cd elctrode. For the real application of the sealed Ni-Zn cells, it is necessary to ionize the evolved O_2 gas from Ni eletrode during charging sealed Ni-Zn cell. Many attempts have been done for enhancing the rate of the O_2 gas ionization by development of cell designs and charging method (17-20).

In the present study, experiments have been performed in sealed Ni-Zn cell to determine the factors which would affect the O_2 gas ionization of Zn electrode. The conditions for improving it during charging sealed Ni-Zn cell have been also found. The effect of the initial State-of-Charge (SOC) of Zn electrode on the O_2 ionization has been investigated in the 25Ah sealed cell. The effect of corrosion inhibitors for the Zn electrode such as HgO, PbO, CdO, In_2O_3 have been also studied. The latter study invetigated the characteristics of the O_2 gas ionization during the cycling.

EXPERIMENTAL

Ni electrodes were prepared by chemical impregnation of $Ni(OH)_2$ active material into sintered Ni plaques which had the porosity of 75%. The impregnated electrodes were cycled in 31% KOH solution with Ni counter electrodes for the formation of the electrodes. Some pairs of the positve electrodes, 70 x 85mm, which had individual theoretical capacity of 4.29Ah were made by spot-welded single electrodes for the assembly of test cells described later. Zn electrodes were made by roll-pressing the mixture of ZnO powder, Zn metal and various corrosion inhibitors with the organic binder PTFE onto Cd coated Cu current collectors in dry state. The roll-pressed Zn electrodes were cut into the size of 70 x 85mm so that each of them could have 11.5Ah of theoretical capacity.

Unwoven polypropylene separator was utilized to wrap Ni electrode and Zn elecrodes individually. Each of Zn electrodes was also wrapped with three layers of microporous polypropylene film (Celgard 3401, Hoechst Celanese, USA) for preventing dendritic Zn growth to the Ni electrode. The sealed cells having nominal capacity of 25Ah were consist of seven Ni electrodes and eight Zn electrodes. The cells were filled with the electrolyte of 6.8M KOH + 1M LiOH under vacuum. The cells were soaked overnight for complete absorption of electrolyte.

All cells were firstly given two formation cycles and then these cells were cycled by charging in three steps and discharging at a rate of C/3 to the voltage of 1.2V. The charge regime of three steps is designed to minimize the O_2 gas evolution from Ni electrodes and also to increase the rate of O_2 gas ionization. Charging at C/5 rate up to 60% SOC and C/20 up to next 25% is followed by charging at C/30 for the last 5%, totally 90% SOC (21).

The pressure transducer (Series 2000, Trans Instruments, England) was connected to the individual sealed cell and continuously monitored the cell pressure developed during the cycling. The pressure measured at the end of second step of charging was used for calculating the absorbed O_2 gas volume.

The anodic polarization curve of Zn electrode was obtained using a Potentiostat (Zahner, IM-5D). The scanning rate was 0.5 mV/sec and the potential was scanned in range of $-1.4 \sim -0.7$ (V vs. Hg/HgO electrode). The porous Zn electrode was used as a working electrode, a Hg/HgO electrode as a reference electrode and a Ni plate as a counter electrode. The Zn working electrode was made by the same process as that for Zn electrodes for the Ni-Zn sealed cells mentioned above. The area of working electrode was designed to be about $1 cm^2$.

RESULTS AND DISCUSSION

Effect of Initial SOC of Zn Electrode on the O_2 Gas Ionization

Table 1. shows the effect of the initial SOC of Zn electrode on absorbed O_2 gas volume during first and second steps of charging. The absorbed O_2 gas volume was determined by assuming that the increase in cell pressure was mainly due to the O_2 gas evolution during the charging and calculating the difference between the theoretically evolved O_2 gas volume until the seond steps of charging and the remaining O_2 gas volume at the end of second step of charging. As the initial SOC of Zn electrode becomes higher, the absorbed O_2 gas volume increase. Since the O_2 gas absorption in sealed Ni-Zn cell is due to the O_2 gas reduction on Zn electrode that is the O_2 gas ionization [1], this result means that the O_2 gas ionization is dependant on the surface area of metallic Zn in Zn electrode.

In order to investigate the O_2 gas ionization quantatively, the current was used for calculating the rate of O_2 gas ionization per time in unit electrode area by applying Faraday's law and the data of absorbed O_2 gas volume from Table 1(22).

$$W = itM/zF \qquad [2]$$
$$I = i/S = zFV_{O2}/5600tS \qquad [3]$$

where, W, the weight of the evolved O_2 gas (g); i, the current used for the O_2 gas ionization (mA); t, the elapsed time (sec.); M, the molecular weight of O_2 gas (g); z, the number of charge involved in the reaction; F, the Faraday constant (96487 C /equivalent); I, the rate of O_2 gas ionization (mA/cm²/sec); S, the total area of Zn electrodes (cm², 700cm² for Ni-Zn 25Ah cell); V_{O2}, the absorbed O_2 gas volume during first and second steps of charging (cm³)

From this relationship, the effect of the initial SOC of Zn electrode on the rate of O_2 gas ionization is revealed and shown in Figure 1. The rate increases linearly as the initial SOC of Zn electrode increases to 40% and then the increase of the rate becomes lower. As mentioned, the rate of O_2 gas ionization is dependant on the surface area of metallic Zn in Zn electrode and this result is accord with the suggestion of McBreen (20).

It was found that the metallic Zn added initially to Zn electrode may improve the electric conductivity of Zn electrode during charging and act as the reaction sites for the O_2 gas ionization. Therefore, the state of metallic Zn surface is one of the most important factors to control the O_2 gas ionization.

Effect of Corrosion Inhibitors for Zn Electrode

With regard to sealed Ni-Zn cell, the H_2 gas evolution on Zn electrode resulting from corrosion has to be suppressed. Corrosion inhibitors with high H_2 gas overvoltage such as HgO and PbO were used in Zn electrode for preventing the H_2 gas evolution. These corrosion inhibitors may affect the condition of metallic Zn surface so that it is necessary to determine the influences of corrosion inhibitors on the O_2 gas ionization.

Figure 2 shows the change in the O_2 gas pressure measured at the end of second step of charging with various corrosion inhibitors during the cycling. Additive compositions including HgO kept the O_2 gas pressure much lower than the other additive such as CdO and PbO. And the O_2 gas pressure became lower as the HgO content increased. Also addition of In_2O_3 and PbO showed relatively lower O_2 gas pressure. As to the case of addition of CdO and PbO, on the other hand, the O_2 gas pressure rapidly increased just after a few cycles and sustained higher O_2 gas pressure during the cycling.

The rate of O_2 gas ionization during the cycling is shown in Figure 3. With addition of HgO and PbO, the rate of O_2 gas ionization maintained higher level of 0.5 ~ 0.9mA/cm^2/sec than that with other additives during the cycling. And addition of In_2O_3 and PbO showed relatively high rate of 0.5 ~ 0.6mA/cm^2/sec through the cycling. Comparing with these additives, the lowest rate of O_2 gas ionization of 0.2 ~ 0.3mA/cm^2/sec with addition of CdO and PbO was obtained. This result clearly indicates that corrosion inhibitors changes the O_2 gas ionization characteristics of Zn electrode and some additives such as HgO, In_2O_3, PbO are effective for improving the O_2 gas ionization during the cycling.

In order to investigate how these corrosion inhibitors for suppressing the H_2 gas evolution make differences in the O_2 gas ionization characteristics, the H_2 gas pressure changes were measured during the operation of the cell and shown in Figure 4. Addition of HgO and PbO lowered the H_2 gas pressure to 1 atm and addition of In_2O_3 and PbO also led H_2 gas pressure as low as 2 atm. As to the case of addition of CdO and PbO, however, the H_2 gas pressure rapidly inicreased to 2.7atm just after 20 cycles and sustainted the high H_2 gas pressure during the cycling. By comparing this result to those obtained from Figure 2 and Figure 3, it can be found that the O_2 gas ionization was motivated more easily under the condition of the low H_2 gas pressure. The accelerated O_2 gas ionization is due to the easy access of O_2 gas to the surface of metallic Zn in the Zn electrode when the cell contains less amounts of H_2 gas.

Corrosion inhibitors also change the passivation behavior of metallic Zn in the Zn electrode, which lead the O_2 gas ionization more easily (22). The curves in Figure 5 show anodic polarization of Zn electrodes with two kind of corrosion inhibitors such as HgO+PbO and CdO+PbO. As the potential moves positively beyond pre-passive region, the dissolution current decreases along with formation of porous ZnO film on the surface of metallic Zn. When the potential reaches to passivation region, the formation of compact ZnO layer ,i.e., the passive film, is initiated on the surface of metallic Zn in Zn electrode (23). Both the pre-passivation potential and the passivation potential with the addition of HgO and PbO were higher than those with the addition of CdO and PbO. And the current density with addition of HgO and PbO was also high within the passivation region. Therefore, addition of HgO and PbO is effective for preventing metallic Zn from passivation so that the surface of the metallic Zn maintains activated and the accelation of the O_2 gas ionization is possible.

By considering the results from Figure 4 and Figure 5, it can be found that the effects of corrosion inhibitors such as HgO, PbO and In_2O_3 on the O_2 gas ionization are due to supressing the H_2 gas evolution and retarding the passivation of metallic Zn.

Characteristics of the O_2 Gas Ionization during the Cycling

With regard to the sealed Ni-Zn cell, the low gas pressure should be kept during the long cycle of sealed cell. It is necessary to investigate the characteristics of the O_2 gas ionization during the cycling. Figure 6 shows the change of the O_2 gas presure and the absorbed O_2 gas volume measured at the end of second step of charging during the operation of the cell. This test was done in the 25Ah sealed cell having Zn electrode with corrosion inhibitor of HgO 1wt% + PbO 0.25wt%. It is appeared that the O_2 gas pressure remained in the range of 0.25 ~ 1.0atm through the cycling. Also the 60 ~ 80% of the evolved O_2 gas volume was absorbed during first and second steps of charging and the absorption of O_2 gas was constantly occurred during the cycling.

The variation of the rate of O_2 gas ionization through the cycling is shown in Figure 7. The rate of ionization was also kept constantly within the range of 0.5 ~ 0.8mA/cm²/sec. It is found that the O_2 gas ionization can be occurred actively on the metallic Zn in the Zn electrode regardless of long operation over 170 cycles.

The sealed Ni-Zn cells with nominal capacity of 3.5Ah were charged to 90% of nominal capacity and discharged on two groups of the cells down to the limit cell voltage of 1.2V and 0.8V respectively. During the cycling, the O_2 gas pressure was monitored to obtain the rate of O_2 gas ionization and shown in Table 2 with Zn electrode potential. When the cell was overdischarged down to 0.8V, Zn electrode potential was under the passive region so that the passivation of metallic Zn became accelated. At this condition, the rate of O_2 gas ionization was decreased 20 ~ 30% of that at the condition of discharge to 1.2V. As discussed in Figure 5, this result indicates that the O_2 ionization was maintained actively due to the preventing surface of the metallic Zn from passivation during the cycling.

Another test was made to confirm the effect of suppressing a passivation during the cycling. The 3.5 Ah seaaled cell which has undergone over 200 cycles was fully discharged down to the cell voltage of 0.15V. From this test, the reserved capacity of Zn electrode was measured. A half of nominal capcity was retained even after 200 cycles and it means that the activeness of metallic Zn surface is well sustained without the passivation throughout the cycling and is consistent with the result shown in Table 2.

Therefore, the O_2 gas ionization constantly occured under the condition which the passivation of Zn electrode was effectively prevented throughout the long operation of the cell.

CONCLUSIONS

The rate of O_2 gas ionization is dependant on the initial State-of-Charge of Zn electrode that is the amount of metallic Zn. This is explained by the facts that metallic Zn promotes the conductivity of Zn electrode and acts as the reaction sites for the O_2 gas ionization.

Corrosion inhibitors for preventing the corrosion of Zn affect the characteristics of O_2 gas ionization on Zn electrode. Especially the additives like HgO, PbO and In_2O_3 are most effective to increase the rate of O_2 gas ionization so that the cell pressure can be maintained lower state during the cycling. This is due to the easier access of O_2 gas to the surface of metallic Zn by suppressing the H_2 gas evolution from Zn electrode and retarding the passivation of metallic Zn throughout the operation of the cell.

The O_2 gas ionization actively occurs on the Zn electrode during the long cycle of the cell under the conditions that the passivation of metallic Zn is prevented.

REFERENCES

1. F. R. McLarnon and E. J. Cairns, J. Electrochem. Soc., **138,** 645 (1991).
2. M. Klein, in Handbook of Batteries and Fuel cells, David Linden, editor, Ch. 19, McGraw-Hill, New York (1984).
3. J. Jindra, J. Power Sources, **37**, 297 (1991).
4. T. S. Lee, J. Electrochem. Soc., **122,** 171 (1975).
5. J. McBreen and E. Gannon, Electrochim Acta, **26**, 1439 (1981).
6. R. A. Jones, U. S. Patent 4,358,517 (1979).
7. H. Tsuji, Jpn. Patent 60,208,053 (1985).
8. L. Binder and W. Odar, J. Power Sources, **13**, 9 (1984).
9. S. Furukawa, K. Inoue and M. Nogami, Jpn. Patent 63,76,270 (1988).
10. O. Wagner and A. Himy, Proceedings of the 27th Power Sources Symposium, PCS Publications Committee, 135 (1976).
11. M. Nogami, K. Inoue and S. Furukawa, Jpn. Patent 63,158,749 (1988).
12. H. Bode, K. Dehmelt and H. von Dohren, Proceeding of the International Symposium on Batteries, Christchurch, Hants, England (1958).
13. E. G. Baars, Proceedings of the 12th Annual Battery Reserch and Development Conference at Fort Monmouth, NJ, 25 (1958).
14. K. Dehmelt and H. von Dohren, Proceedings of the 13th Annual Battery Research and Development Conference at Fort Monmouth, NJ, 85 (1959).
15. D. R. Turner, Electrochem. Technol., **2**, 313 (1964).
16. J. Jindra and M. Musilova, Abstract P:III - 1 - 17 in the 44th International Society of Electrochemistry Meeting Abstracts, Berlin, Germany (1993).
17. M. Kanematsu, Jpn. Patent 61,78,058 (1986).
18. K. Fujii, M. Yamane and T. Ueda, Jpn. Patent 01,100,872 (1989).
19. D. P. Boden and E. Pearlman, in Power sources vol. 4, D. H. Collins, editor, p. 103, Oriel press, Newcastle-upon-Tyne (1973).
20. J. McBreen, Proc. Electrochem. Soc., 77, 724 (1977).
21 Y. B. Pyun, H. Chang and B. H. Jung, Abstract P:1 - 17 in the 46th International Society of Electrochemistry Meeting, Xiamen, China (1995).
22. H. Chang, Y. B. Pyun and M. E. Alexeeva, The 188th Electrochemical

Society Meeting, Chicago (1995) in print.
23. M. B. Liu, G. M. Cook and N. P. Yao, J. Electrochem. Soc., **128,** 1663 (1981).

Table 1. Absorbed O_2 gas volume during first and second steps of charging at various initial State-of-Charge of Zn electrode in sealed Ni-Zn cell

The initial SOC of Zn Electrode (%) [a]	0	10	20	30	40	50	60
Absorbed O_2 gas volume (cm^3)[b]	12	31	45	70	86	89	92

a : The weight fraction of total active materials of Zn Electrode
b : Determined at P_{O2} = 0.5 atm

Table 2. Effect of Zn electrode potential on the O_2 gas ionization (Nominal Capacity : 3.5Ah)

Zn electrode potential (V vs. Hg/HgO)	Rate of O_2 gas ionization ($mA/cm^2/sec$)
- 1.0 ~ - 0.5 (pre passive & passive region)	0.11 ~ 0.18
- 1.1 ~ - 1.2 (dissolution region)	0.5 ~ 0.6

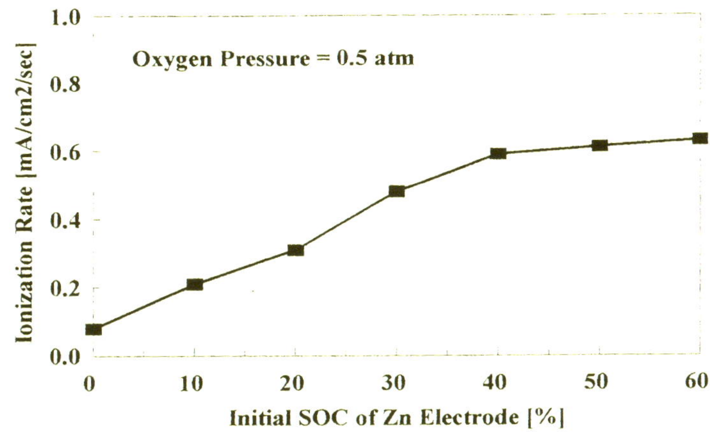

Figure 1. Effect of the initial SOC of Zn electrode on the rate of O_2 gas ionization in sealed 25Ah Ni-Zn cell

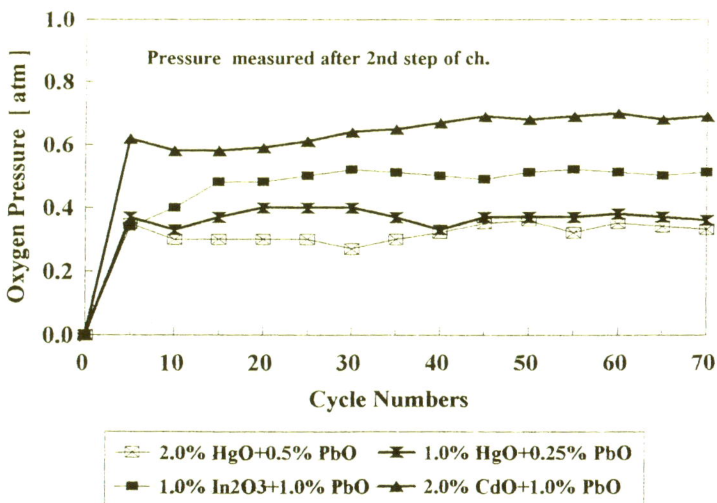

Figure 2. Effect of corrosion inhibitors on the O_2 gas pressure in sealed 25Ah Ni-Zn cell

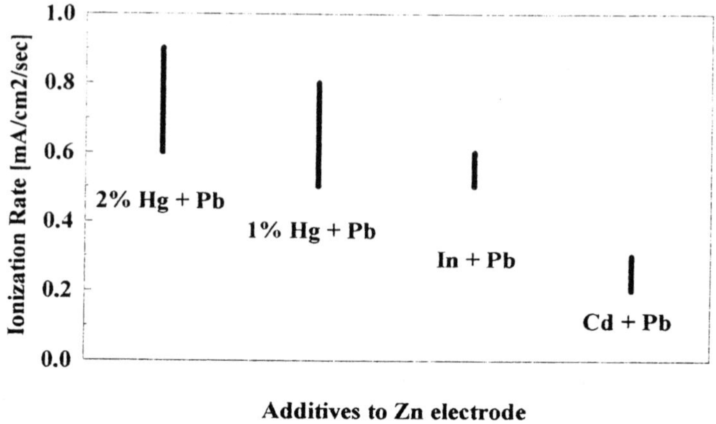

Figure 3. Effect of corrosion inhibitors on the rate of O_2 gas ionization in sealed 25Ah Ni-Zn cell

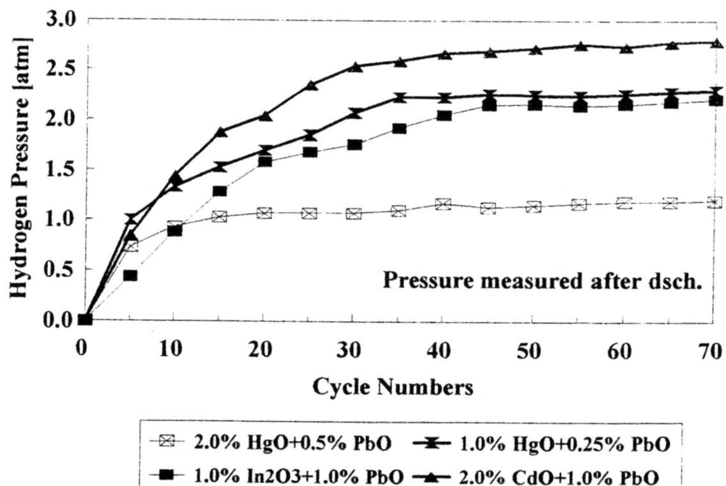

Figure 4. Effect of corrosion inhibitors on the H_2 gas evolution in sealed 25Ah Ni-Zn cell

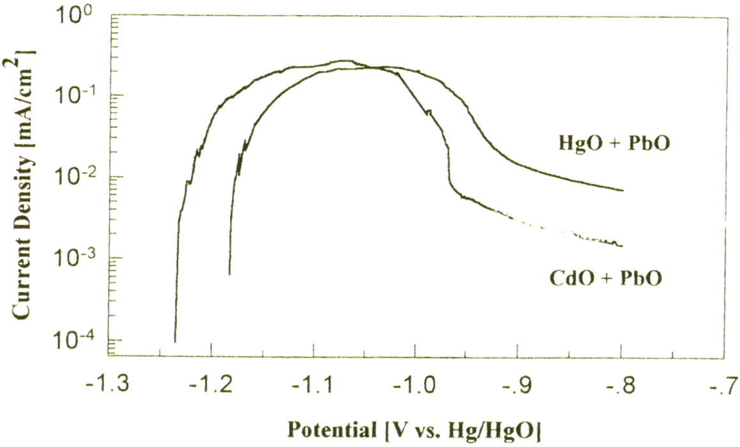

Figure 5. Anodic polarization curves of Zn electrode with different corrosion inhibitors

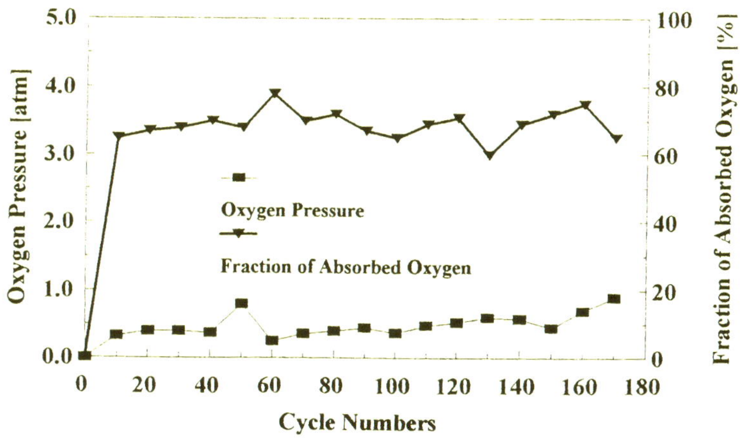

Figure 6. Changes of the O_2 gas pressure and the absorbed O_2 gas volume during the cycling in sealed 25Ah Ni-Zn cell

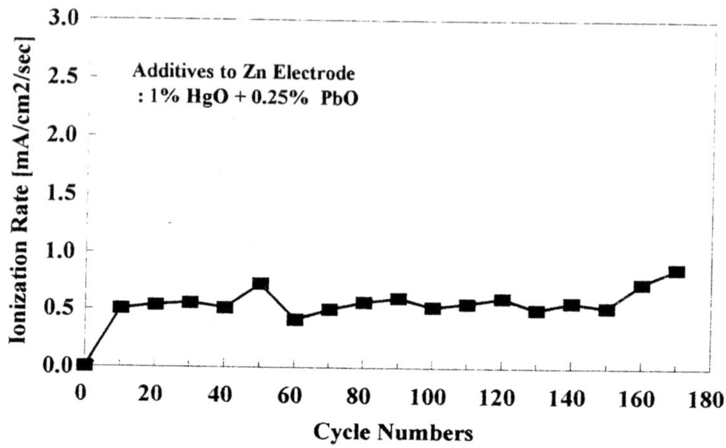

Figure 7. Change of the rate of O_2 gas ionization during the cycling in sealed 25Ah Ni-Zn cell

FACTORS INFLUENCING THE ELECTROCHEMICAL DEPOSITION OF NICKEL HYDROXIDE FILMS

C.C. Streinz[1], S. Motupally and J.W. Weidner
Center for Electrochemical Engineering
Department of Chemical Engineering
University of South Carolina
Columbia, SC 29208

An electrochemical quartz crystal nanobalance (EQCN) was used to measure the mass of nickel hydroxide films electrochemically deposited from nickel nitrate solutions. The objective was to quantify deposition as a function of the following deposition conditions; applied current, nickel nitrate concentration, temperature and ethanol content in the solvent. The concentration of nickel nitrate was found to have the greatest effect on deposition efficiency, with the efficiency increasing by more than one order of magnitude with a reduction in concentration from 2.0 to 0.2 M.

Introduction:

Nickel hydroxide is one of the most commonly used active materials for positive electrodes in secondary batteries, such as NiZn, NiCd and NiH. Nickel hydroxide is commonly prepared electrochemically by cathodic precipitation from a nickel nitrate bath. It has been shown that the electrochemical impregnation of porous nickel plaques produces superior battery electrodes compared to those made by the conventional loading process (see Gross [1] for a review of both the chemical and electrochemical impregnation processes). In electrochemical impregnation, a nickel plaque is cathodically polarized in a nickel nitrate solution. Nitrate is reduced according to the following reaction [2-6].

$$NO_3^- + 6H_2O + 8e^- \Rightarrow NH_3 + 9OH^- \qquad (1)$$

One additional OH^- ion per eight electrons passed may be generated from the ammonia/ammonium-ion equilibrium as expressed in reaction 2.

---

[1] Present Address: Cabot Corporation, Microelectronics Materials Division, 500 Commons Drive, Aurora, IL 60504

$$NH_3 + H_2O = NH_4^+ + OH^- \quad (2)$$

In acid and neutral conditions equilibrium tends towards the right-hand side of equation 2 [7], and therefore, if the pH at the electrode surface does not exceed 8, the nitrate reduction reaction can effectively be written as follows [8-10].

$$NO_3^- + 7H_2O + 8e^- \Rightarrow NH_4^+ + 10OH^- \quad (3)$$

The production of OH^- increases the local pH, resulting in the precipitation of $Ni(OH)_2$ according to reaction 4 [3-6,8-10].

$$Ni^{++} + 2OH^- \Rightarrow Ni(OH)_2\downarrow \quad (4)$$

The objective of this study was to quantify deposition as a function of the following deposition conditions; applied current, $Ni(NO_3)_2$ concentration, temperature and ethanol content in the solvent. The primary motivation for quantifying the deposition process with respect to the above parameters is to gain a more fundamental understanding of the deposition process in order that the impregnation of porous plaques may be more thoroughly and fundamentally modeled [5,11].

Experimental:

An Electrochemical Quartz Crystal Nanobalance (EQCN - Elchema model EQCN-501) was utilized to make *in-situ* measurements of the mass deposited on a planar gold electrode during cathodic chronopotentiometry in nickel nitrate solutions. It has previously been demonstrated that the measured mass change results from the deposition of $Ni(OH)_2$ rather than metallic Ni [12,13]. An illustration of the EQCN cell configuration can be seen in Figure 1. As shown, the working electrode consists of a quartz crystal with a thin layer of gold sputtered on both sides. The electrode area exposed to solution during deposition is 0.2 cm^2. The resonant frequency of the reference crystal is 10.000 MHz. The EQCN has a frequency resolution of 1 Hz which corresponds to 1.1 ng mass change [14]. A Pine Model AFRDE5 bi-potentiostat was used for all current and potential control.

Depositions were carried out in $Ni(NO_3)_2$ solutions, whose concentrations ranged from 0.01 - 3.0 M at applied current values ranging from 0.01-1.0 mA. Solvents consisting of 0, 50, 75 and 100 v/o ethanol were used [it should be noted that the solution consisting of 100 v/o ethanol was not truly anhydrous since the solute contained a 6:1 molar ratio of H_2O:$Ni(NO_3)_2$]. Depositions were carried out at room temperature (23-25°C), and controlled temperatures of 50 and 75°C using a jacketed flow through cell.

The solution pH was found to be a function of $Ni(NO_3)_2$ concentration, ranging from 2.4 for 2.0 M solution to 4.0 for 0.1 M solution. In order to determine the effect of pH on deposition rate a solution containing 0.2 M $Ni(NO_3)_2$ was acidified to solution pH's of 3.0, 2.5, 2.0 and 1.5 using nitric acid. In addition, a solution containing 2.0 M $Ni(NO_3)_2$ was basified to pH 4.0 using concentrated KOH. Depositions were carried out in these solutions at an applied current of 0.5 mA and the data compared to that for the non-acidified or basified solutions. Additional depositions were carried out in solutions saturated in NO_3^- (obtained by adding excess $NaNO_3$) to determine the role of NO_3^- in deposition.

Films deposited under selected conditions were rinsed in distilled water, charged and then discharged at a constant current of 50 μA in 3.0 w/o KOH. All discharges were conducted at room temperature. The electrochemical capacity of each film (determined by the time to discharge) was compared to its deposited mass and used to evaluate the deposits obtained under the different deposition conditions.

Results and Discussion:

The Effect of Time and Current: Figure 2 is a plot of mass vs. time for deposition from 0.2M $Ni(NO_3)_2$ at applied currents of 0.05, 0.1, 0.25 and 0.5 mA (0.25, 0.5, 1.25 and 2.5 mA/cm^2). The mass vs. time curves given in this paper are reproducible within a relative error of 5%. The mass of the deposit increased linearly with time (with regression values, R^2, consistently greater than 0.999) indicating a constant rate of deposition. Linearly increasing mass was observed at all deposition conditions with the exception of very low solution concentrations at high currents (e.g. 0.01M @ 0.5 mA). The effect of applied current can also be observed in Figure 2. Increasing current was observed to increase the deposition rate as indicated by the increasing slopes in the figure.

The effect of $Ni(NO)_2$ Concentration: The effect of solution concentration on the deposition rate is illustrated in Figure 3, which plots mass vs. deposition time in several solution concentrations at 0.5 mA (2.5 mA/cm^2). As discussed previously the mass is observed to increase linearly with time at each solution concentration. Note, however, that the rate of deposition (slope of the mass-time curve) increases with decreasing $Ni(NO_3)_2$ concentration. The deposition rates at each concentration are indicated on the figure. Similar behavior (increasing rate with decreasing concentration) was observed at other current values (0.1, 0.25 and 1.0 mA).

The relationship between deposition rate at an applied current of 0.5 mA and $Ni(NO_3)_2$ concentration is illustrated in Figure 4, which plots the deposition rate with respect to the inverse of concentration. The horizontal dashed lines show theoretical deposition rates at which 100% of the electrochemically produced OH^- is utilized in the deposition of $Ni(OH)_2$. According to Faraday's Law, the horizontal line is given as follows

$$\frac{dm}{dt} = \frac{S_{OH^-} iAM^o_{Ni(OH)_2}}{2F} \tag{5}$$

where *dm/dt* is the deposition rate (g/s), S_{OH}^- is the stoichiometric coefficient for hydroxyl ions per electron (from Eqns 1 and 3), i is the applied current density (A/cm^2), A is the electrode area ($0.2\ cm^2$), and $M^o_{Ni(OH)_2}$ is the molecular weight of pure $Ni(OH)_2$ (93 g/mol). The two in the denominator results from needing two mols of hydroxyl per mol of deposited $Ni(OH)_2$. Note that Eqn 3 predicts that mass increases linearly with time and that mass is independent of solution concentration. It can be seen in Figure 4 that at low $Ni(NO_3)_2$ concentrations (0.2 and especially 0.1 M) nearly 100% of the generated OH^- is consumed in the deposition of $Ni(OH)_2$. However, at higher $Ni(NO_3)_2$ concentrations the efficiency of OH^- utilization decreases dramatically (to less than 20% at solution concentrations greater than 1.0 M). Further, at high concentrations the rate of deposition is shown to be linearly related to the inverse of the solution concentration (with a regression value, R^2, of 0.997).

This type of behavior was observed over a range of one order of magnitude of current (0.1 - 1.0 mA). The data at different applied currents can be normalized by plotting the utilization efficiency of the electrochemically generated OH^- vs. inverse concentration as illustrated in Figure 5. Note that the efficiency of utilization, while increasing slightly with current, is relatively independent of the deposition current. At all currents the utilization is highly inefficient in concentrated solutions and nearly 100% efficient in dilute solutions.

It was shown previously [12] that the increased NO_3^- concentration and the decreased pH in concentrated $Ni(NO_3)_2$ solutions are not the primary causes of the reduced rate of deposition. The mechanism must therefore be related to the increased Ni(II) concentration in solution. Nickel is well known to complex in aqueous solutions, forming various soluble species (e.g., $NiOH^+$, $Ni(OH)_2(aq)$, $Ni(OH)_3^-$ and $Ni(OH)_4^{2-}$ [15]). Perrin [16] demonstrated that Ni^{++} is in equilibrium with $NiOH^+$ in dilute solutions. Other researchers (see references in Baes and Mesmer [15]) have shown that the predominant species in more concentrated Ni(II) solutions (> 0.015 M) is the polymeric $Ni_4(OH)_4^{4+}$. Assuming the complex is $Ni_4(OH)_4^{4+}$, a two step precipitation reaction mechanism has been proposed [12]:

$$4Ni^{2+} + 4OH^- \Leftrightarrow Ni_4(OH)_4^{4+} \tag{6}$$

$$Ni_4(OH)_4^{4+} + 4OH^- \Rightarrow 4Ni(OH)_2\downarrow \tag{7}$$

Depending on the deposition conditions, $Ni_4(OH)_4^{4+}$ either reacts with the OH^- and deposits as $Ni(OH)_2$, according to Eqn. 7, or it diffuses away from the electrode. If there is significant diffusion away from the electrode, deposition rates will be less than that predicted by Faraday's law. Since the concentration of $Ni_4(OH)_4^{4+}$ is essentially equal to zero in the acidic environment of the bulk solution [12, 15], the concentration gradient of $Ni_4(OH)_4^{4+}$ is proportional to its concentration near the electrode surface. In concentrated $Ni(NO_3)_2$ solutions, where the concentration of $Ni_4(OH)_4^{4+}$ is high at the electrode surface [12], the diffusion of $Ni_4(OH)_4^{4+}$ away from electrode will be rapid. This process consumes OH^- and decreases the efficiency of utilization of electrochemically generated OH^-. Equilibrium calculations indicate that decreasing the concentration of $Ni(NO_3)_2$ from 1.0 to 0.1 M decreases the concentration of $Ni_4(OH)_4^{4+}$ at the electrode surface by a factor of 15 (from 0.20 to 0.013 M) and hence the diffusion of this species. The net result is an increase in the rate of deposition by a factor of 5 (see Table 1 in ref [12]).

The Effect of Temperature: The effect of temperature is illustrated in Figure 6, which is a plot of mass *vs.* time for deposition in 1.0M $Ni(NO_3)_2$ dissolved in a solvent mixture of 50/50 v/o ethanol/water. Independent of the temperature, the mass is observed to increase linearly with time (with regression values, R^2, consistently greater than 0.999), indicating a constant deposition rate. The deposition rates (slopes of the mass *vs.* time curves) are essentially identical at RT and 50°C. However, the deposition rate at 75°C, is approximately 65% greater than that at RT and 50°C (5.6 *vs.* 3.4 μg/min). Also shown in Figure 1 are the theoretical mass *vs.* time curves (for reduction of nitrate according to Eqns. 1 and 3) in which the slopes are based on Faraday's law, which is given by Eqn 5.

Figures 7a and 7b illustrate the effect of $Ni(NO_3)_2$ concentration on the deposition rates at RT, 50 and 75°C. The horizontal dashed lines show the theoretical deposition rates (calculated from Eqn. 5) for the nitrate reduction reactions given in Eqns. 1 and 3. Figure 7a is a plot of deposition rate *vs.* inverse solute concentration in 50/50 v/o ethanol/water solvents. At each concentration (over a range of 1.0 to 0.1 M) the deposition rates were virtually identical at RT and 50°C, while the deposition rate at 75°C is up to 70% greater than that observed at the lower temperatures (13.6 *vs.* 8.0 μg/min at 0.5 M). Figure 7b is a similar plot to Figure 7a but for solutions with aqueous solvents. At all concentrations (with the exception of 0.1M) the deposition rate increases with increasing temperature with the most significant increase occurring between 50 and 75°C. At 0.1 M the trend observed is identical to that observed in Figure 7a (*i.e.*, similar deposition rates at RT and 50°C and a significant increase in rate at 75°C).

Figures 7a and 7b show that under certain conditions (0.2 and 0.1 M $Ni(NO_3)_2$ at 75°C) the deposition rate is significantly greater than that predicted by Faraday's law (Eqn. 5). There are two possible explanations; first, the electrochemical reaction for the reduction of nitrate could be different than that given in Eqns. 1 or 3 and second, the molecular mass of the deposited film may be greater than that of pure $Ni(OH)_2$. Eqn. 3 gives the highest stoichiometric constant reported (10:8) [1], yet is not sufficient to explain the high deposition rates observed. It is more likely that the molecular mass of the deposition product is greater than 93 g/mol, indicating that the deposited films contain co-deposited water and/or ethanol. This is consistent with the widely accepted view that the deposition product is α-$Ni(OH)_2$, which has been shown to contain approximately $^2/_3$

mol of H_2O per mol $Ni(OH)_2$ [17,18]. The effect of deposition conditions on the atomic mass of the deposition product is discussed in greater detail elsewhere [13].

The Effect of Ethanol: A plot of mass versus time for deposition in 1.0 M $Ni(NO_3)_2$ dissolved in solvents consisting of 0, 50, 75 and 100 v/o ethanol at 0.5 mA (2.5 mA/cm^2) is shown in Figure 8. Independent of the solvent composition, the mass of the deposit increased linearly with time, indicating a constant rate of deposition. Deposition rates were observed to increase as the ethanol content increased, with the most significant increase observed between 75 and 100 v/o. The measured deposition rates were 2.7, 3.4, 3.9 and 8.6 μg/min for the 0, 50, 75 and 100 v/o ethanol solutions, respectively. Also plotted in Figure 5 are the mass *vs.* time curves predicted from Faraday's law (Eqns. 1, 3, and 5 - see above).

The effect of $Ni(NO_3)_2$ concentration on the deposition rate at a current of 0.5 mA is illustrated in Figure 9 which is a plot of deposition rate *vs.* inverse $Ni(NO_3)_2$ concentration. The horizontal dashed lines show the theoretical deposition rates calculated from Eqn. 5. At both 1.0 and 0.5 M $Ni(NO_3)_2$ the deposition rate increases with increasing ethanol content. The shapes of the three curves for the solutions containing 0, 50 and 75 v/o EtOH are similar while the deposition rate for the 100% ethanol increases more gradually with decreasing concentration. As a result, at a concentration of 0.1 M the slowest rate is observed in the 100% ethanol solution

Summary and Conclusions:

The results of this experimental study can be summarized by illustrating five main trends in the data [12,13]: (1) the mass of the deposited $Ni(OH)_2$ increased linearly with time at all deposition conditions; (2) the deposition rate (mass per unit time) was approximately proportional to the applied current; (3) the deposition rate increased with decreasing $Ni(NO_3)_2$ concentration ; (4) the deposition rate increased by approximately 50% when the bath temperature was increased from 25 to 75 °C; and (5) the deposition rate increased with increasing ethanol content in the solvent for concentrated $Ni(NO_3)_2$ solutions.

The fundamental understanding of the deposition process gained through this work should enable the impregnation of porous plaques to be more thoroughly modeled and ultimately enable higher quality and less expensive electrodes to be manufactured. These data are also being used in conjunction with a mathemeatical model of deposition on a planar electrode to deduce the chemical equilibrium constants that govern the process [19].

References:

[1] S. Gross, "Review of Electrochemical Impregnation for Nickel Cadmium Cells," Jet Propulsion Laboratory, California Institute of Technology, Pasadena, CA (1977).

[2] H.N. Seiger and V.J. Puglisi, Proceedings of the Power Sources Symposium, **26**, 115 (1974).

[3] D.F. Pickett and J.T. Maloy, *J. Electrochem. Soc.*, **125**, 1026 (1978).

[4] K. Ho, *J. Electrochem. Soc.*, **134**, 52C (1987).

[5] K. Ho and J. Jorne, *J. Electrochem. Soc.*, **137**, 149 (1990).

[6] F. Portemer, A. Delahaye-Vidal, and M. Figlarz, *J. Electrochem. Soc.*, **139**, 671 (1992).

[7] J.A. Dean, editor, Langes Handbook of Chemistry, 14th Ed., McGraw-Hill, Inc., New York (1992).

[8] P.K. Ng and E.W. Schneider, *J. Electrochem. Soc.*, **133**, 17 (1986).

[9] D.A. Corrigan and R.M. Bendert, *J. Electrochem. Soc.*, **136**, 723 (1989).

[10] J. Desilvestro, D.A. Corrigan and M.J. Weaver, *J. Electrochem. Soc.*, **135**, 885 (1988).

[11] G. Nagarajan, C.H. Ho and J.W. Van Zee, Extended Abstract 60, Electrochemical Society Extended Abstracts, Pennington, NJ, October 9-14 (1994).

[12] C.C. Streinz, A.P. Hartman, S. Motupally and J.W. Weidner, *J. Electrochem. Soc.*, **142**, 1084 (1995).

[13] C.C. Streinz, S. Motupally and J.W. Weidner, *J. Electrochem. Soc.*, In Press (1995).

[14] Technical Manual, Electrochemical Quartz Crystal Nanobalance, System EQCN-500, Elchema, Potsdam, NY.

[15] C.F. Baes and R.E. Mesmer, The Hydrolysis of Cations, John Wiley and Sons, New York, 242-247 (1976).

[16] D.D. Perrin, *J. Chem. Soc.*, 3644 (1964).

[17] H. Bode, K. Dehmelt and J. Witte, *Electrochim. Acta*, **11**, 1079 (1966).

[18] D.M. MacArthur, *J. Electrochem. Soc.*, **117**, 422 (1970).

[19] M. Murthy, G. Nagarajan, J.W. Weidner and J.W. Van Zee, submitted to *J. Electrochem. Soc.* (1995).

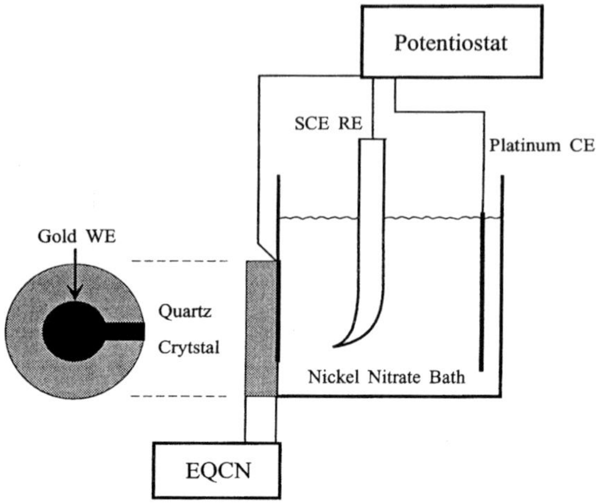

Figure 1: An illustration of the EQCN cell configuration. The working electrode consists of a quartz crystal, sputtered with a thin layer of gold on both sides. The electrode area exposed to solution is 0.2 cm^2.

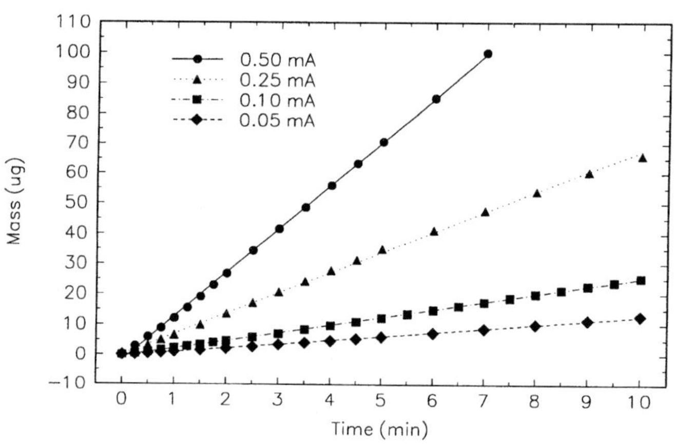

Figure 2: A plot of mass vs. time for deposition in 0.2M $Ni(NO_3)_2$ at applied currents of 0.05, 0.1, 0.25 and 0.5 mA (0.25, 0.5, 1.25 and 2.5 mA/cm^2) Note the linearly increasing mass with deposition time.

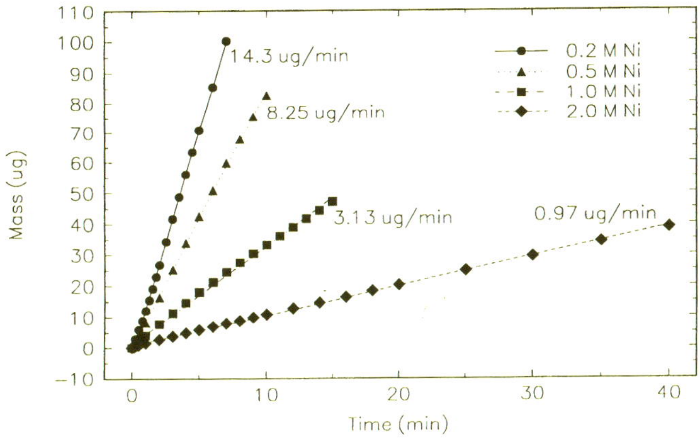

Figure 3: A plot of mass vs. time at 0.5 mA in various $Ni(NO_3)_2$ concentrations. Note that the rate of deposition increases dramatically with decreasing solution concentration. The deposition rates at each concentration are indicated on the figure.

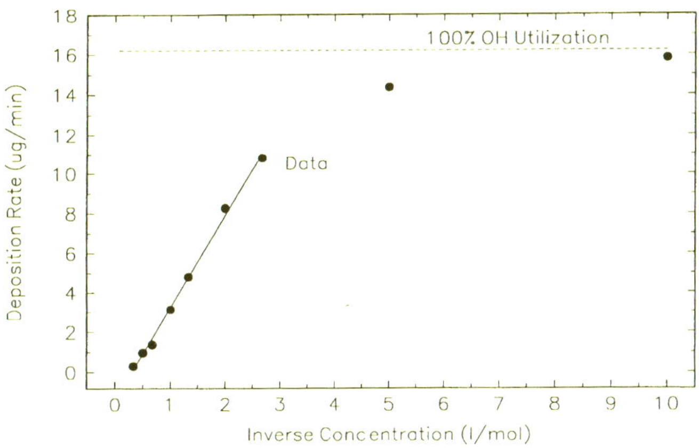

Figure 4: A plot of deposition rate at 0.5 mA vs. inverse $Ni(NO_3)_2$ concentration. The horizontal dashed line shows a theoretical deposition rate predicted by Faraday's Law. The deposition rate is observed to be both highly inefficient and linearly related to inverse concentration in concentrated solutions.

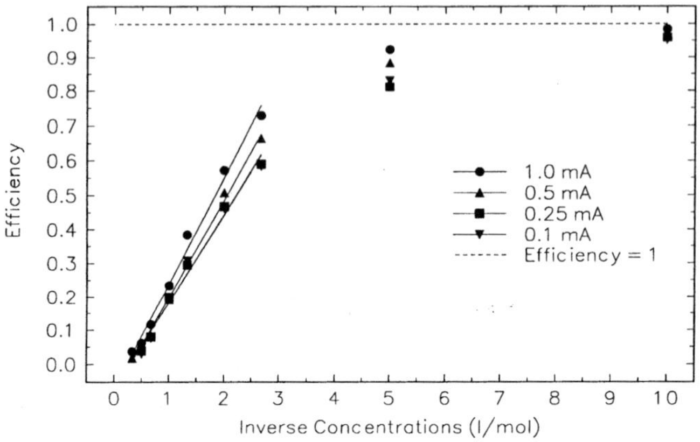

Figure 5: A plot of the utilization efficiency of electrochemically generated OH^- vs. inverse $Ni(NO_3)_2$ concentration. The efficiency is relatively independent of current. At all currents utilization of OH^- is highly inefficient in concentrated solutions and nearly 100% efficient in dilute solutions.

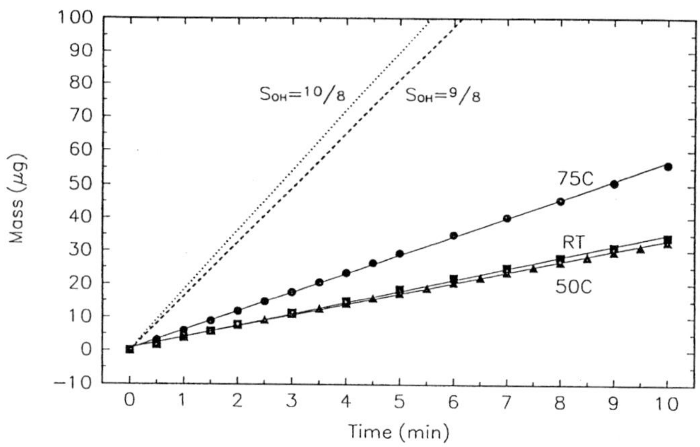

Figure 6: Mass *vs.* time for deposition in 1.0M $Ni(NO_3)_2$ at 0.5 mA in 50/50 ethanol/water solutions at RT, 50 and 75°C. The deposition rate at 75°C, is approximately 65% greater than that at RT and 50°C. The dashed lines are the theoretical mass *vs.* time curves from Eqn. 5 for the two values of S_{OH^-}.

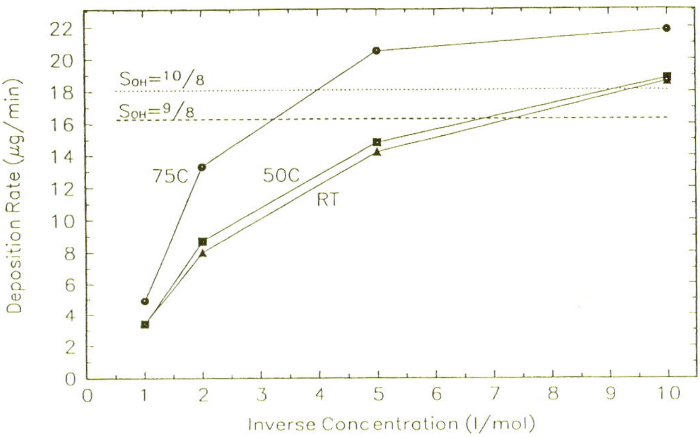

Figure 7a: Deposition rate *vs.* inverse concentration in 50 v/o ethanol solvents at RT, 50 and 75°C. The dashed lines are the theoretical rates from Eqn. 5 for the two values of S_{OH^-}. A significant increase in deposition rate is observed at 75°C at all concentrations.

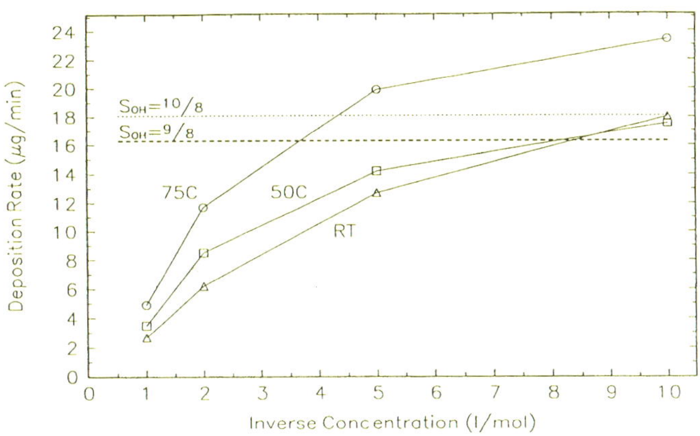

Figure 7b: Deposition rate *vs.* inverse concentration in aqueous solvents at RT, 50 and 75°C. The dashed lines are the theoretical rates from Eqn. 5 for the two values of S_{OH^-}. A significant increase in deposition rate is observed at 75°C at all concentrations.

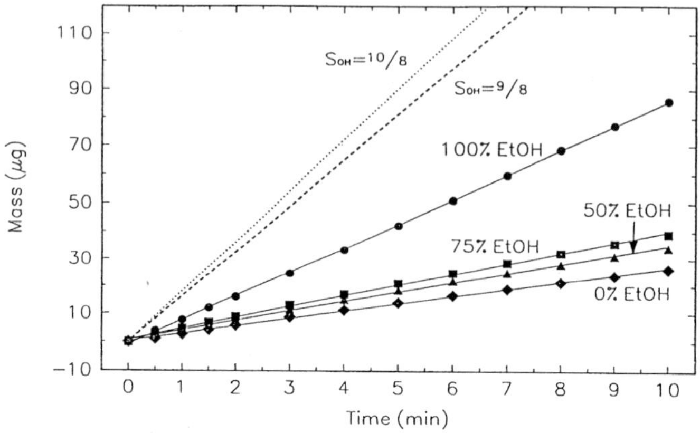

Figure 8: Mass *vs*. deposition time for deposition in 1.0M $Ni(NO_3)_2$ at 0.5 mA. The measured deposition rates were 2.7, 3.4, 3.9 and 8.6 μg/min for the 0, 50, 75 and 100 v/o ethanol solutions, respectively. The dashed lines are the theoretical mass *vs*. time curves from Eqn. 5 for the two values of S_{OH^-}.

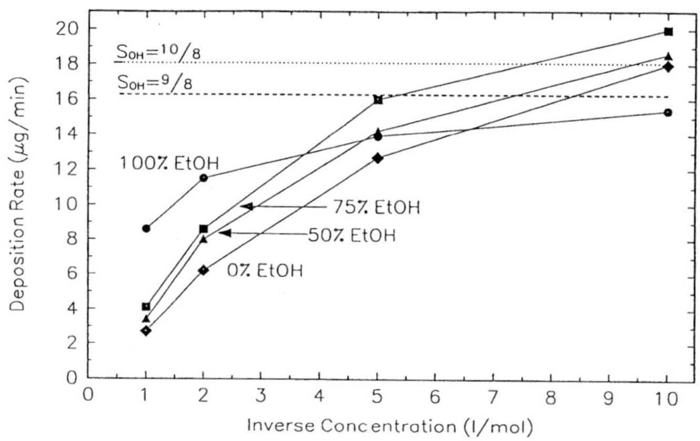

Figure 9: Deposition rate *vs*. inverse $Ni(NO_3)_2$ concentration at 0.5 mA for 0, 50, 75 and 100 v/o ethanol solutions. The horizontal dashed lines show the theoretical deposition rates calculated from Eqn. 5 for the two values of S_{OH^-}.

The Palladium Medal Award - A background discussion

A. J. Salkind
UMDNJ- Robert Wood Johnson Medical School
and Rutgers University
Piscataway, NJ 08855

Prof Aleksandr Naumovich Frumkin was the fifth person to receive the Palladium Medal of the Electrochemical Society. This award in 1959 continued the line of internationally distinguished recipients. The earlier Palladium medals, the highest recognition by the society, had been awarded as follows:

C.W.Wagner (1951)
N.H.Furman (1953)
U.R.Evans (1955)
K.F.Bonhoeffer (1957)

At the time of the award, the following announcement appeared in the Journal of the Electrochemical Society;

"Prof Frumkin is a member of the Academy of Sciences of The USSR, Professor of Electrochemistry at the University of Moscow (a Chair held for twentyfive years), and Director of the Electrochemical Institute of the USSR. He is undisputely the leading figure in electrochemistry at the present time and his influence is felt in practically all fields associated with interfaces....

Frumkin's earliest work (about 1916) concerned the theory of adsorbed layers and equations of state for these films. His most well known early paper is probably that on the effect of the electrokinetic potential on the concentration of hydroxonium ions in the Helmholtz double layer. He introduced, in this way, the relation between the structure of the double layer and the kinetics of reactions... The most extensive studies carried out by Frumkin relate to that central electrochemical reaction, the hydrogen evolution reaction, in which the rate controlling proton discharge on mercury from aqueous solutions was established...

The theory of mixed electrodes was examined as early as 1941, and these studies contributed much to the understanding of current distribution at corroding materials...

Frumkin inspired the creation of a great outpouring of electrochemical studies from the Soviet Union, which have multiplied as the large number of co-workers who passed through his group fructified the work of other groups, both in research institutes and industry. In this connection, leading contributions have been made to the fields of mass transport, corrosion inhibition, hydrogen embrittlement, wetting, flotation and batteries. Of the many electrochemists of international repute who have emanated from Frumkin's laboratory, a few of those best known in the West may be named as: Bagotzky, Burshtein, Erschler, Dolin, Jofe, Kabanov, Levich, Levina, Lukovstev, Slygin, and Tempkin."

Frumkin's distinguished career included the following events of note:

President of CITCE (Now the International Electrochemical Society) 1965.

Member of the Polish Academy of Science, National Academy of Sciences of the USA, Akademie Deutscher Naturforscher Leopoldina, Bulgarian Academy of Sciences, Royal Academy of Sciences of the Netherlands, the American Academy of Sciences and Arts, Honorary doctorates from the Universities of Dresden and Brno as well as the many awards and citations given to him by the Russian (Soviet) government.

Although he was best known for his fundamental contributions to electrochemical and physical chemical phenomena, Frumkin was also concerned with many practical applications. These included fuel cells, bioelectrochemistry, the design and construction of electrochemical instruments, the electrochemical machining of metals, and semiconductors.

I first met Frumkin during his visits to the Bockris laboratories at the University of Pennsylvania in 1959 and 1960. His lectures were incisive, clear, and in very good English. I was then employed nearby at the research center of the Electric Storage Battery Company in Yardley Pennsylvania. Subsequently, at a meeting of CITCE in London, I was seeking Prof. Levich in order to clarify some aspects of rotating disc experiments I was carrying out. Levich was not in sight but Frumkin was standing at the entrance to the conference hall. I told him of the help I was seeking and asked him if he knew Dr. Levich's whereabouts . His reply to my query was a question; "Do you want theoretical or practical information and discussion". I indicated that I wanted some of both, but that the practical was of greater importance to me. Frumkin then stated, " then you don't necessarily want Levich, you want me". With much clarity, Frumkin was able to explain most of the minute details of the great work of Levich.

THE GREAT ACADEMICIAN- A REMINISCENCE *

John O'M. Bockris
Department of Chemistry
Texas A & M University
College Station, TX 77843

My first personal impression of the 'Great Man' was a hurried handshake at the steps of the plane in Moscow Airport in 1956. I was taken aback by the fact that he had come to meet me (and continued to be surprised by that fact every time I went to Moscow in the next 20 years) but there he was, at the bottom of the airplane's steps. In later years, he came only to the edge of the tarmac. No customs formalities, of course-he flashed his Academy card- and in three or four minutes after stepping out of the plane, we were at the great black, chauffeured limousine (a provision from the Academy) outside the airport.

Frumkin would never ask about the flight or other 'nonsensical' trivia. As we slide into the back seats, he said sharply: "Good. Now, I want to discuss your last paper".

That is how it began, and that is how it went on for the next 20 years, with 5 visits and letter exchanges about every two months. His knowledge was encyclopedic, his acumen extreme, and his ability to recall the literature from the past, and to combine it with an absolutely up-to-date knowledge of electrochemistry, constituted a great part of his success.

Frumkin remained throughout his life, even in the 1970s, what might be called the 'God Professor'. It may well have been that formally he was responsible with government funding committees. Within the Institute for Electrochemistry his power was supreme. He possessed superior intellect and a very high energy level. He had remarkable ability to discuss clearly scientific work within his field. His power of organization was good, too, and he showed in later years 'management ability' of the carrot variety- he arranged the award of many prizes to his co-workers. He was primarily an intellectual. Compared to other professors, he showed a devastingly wide knowledge of the modern humanistic literature, and was a connoisseur of e.g. old churches, and botanical aspects of the local scene. He read widely in French, German, and English and it was an experience to hear him, perhaps at a dinner in his dascha, take up now with this French professor, now with that German professor, now with this American professor, a vigorous discussion of the latest in literature of the various countries, in their own language.

In this great range of intellectual knowledge, he was joined by the second of his three wives, Obruchewa, an electrochemist, who also had a corresponding knowedge of the humanistic literature. A leading scientist was expected to be an intellectual who applied himself to science as part of knowledge. Frumkin had plenty of vitality, as his marriage to a third wife at the age of 72 (" well, she is a professor") illustrated. Few men who knew him well were surprised at his remarriage

What did he really achieve?

I have sometimes found it difficult to list contributions of long-lasting and world stature to back up the extremely high reputation with which Frumkin's name (and thus Russian electrochemistry) is associated. One of the reasons for this is that there is so much to choose from.

Firstly, of course, there is the fantastic number of publications. There must be more than one thousand, although he often wanted to diminish it to a few hundred of the best, as in the list of 400 selected papers he gave me when I proposed him for the Nobel prize in 1966.

Secondly, there was his strong advocacy of electrochemistry, his zest for the subject, his productivity of over 50 years, and the stream of high standard work.

Thirdly, the spirit of the man. his personality and dynamicism.

What are Frumkin's most important Electrochemical contributions of lasting value:

1- His laboratory showed how to do electrode kinetic measurements, free from solution impurity effects at very low rates, when tiny traces of impurities can compete with the main reactions.

2- The use of ac work in analysing processes at interfaces was published 5 years before the usually quoted work of Randalls.

3- He was the originator of an overwhelmingly large amount of the data on organic absorption on mercury, and did for this filed what Grahame had done in America for the corresponding inorganic adsorption.

4- Frumkin gave the early and more elegant treatments of the kinetics of consecutive reactions. In these, along with Horiuti, made original contributions to the concept of rate-determining step.

5- He introduced the concept of the potential of zero charge at electrodes, although it was later taken over and emphasized by Antropov.

6-He showed(with Petry and others) how to obtain the components of charge at the metal-solution interface of non-polarizable electrodes.

However, others in the electrochemistry laboratory of the Academy produced magnificently, and we must here acknowledge the work of Levich, in his famous contributions to electrochemical hydrodynamics; Bagotsky, in his elegant elucidations of the kinetics of organic reactions; and a host of others, for example Brodski (deceased). Damaskin, Filitovski, Kazarinov, Petry, Pleskov, Ninas, Schumilowa, and Osetrowa. As the very active scientific, as well as administrative, head of the Institute of Electrochemistry of the Academy of Sciences, Frumkin contributed to the work of all these people, whilst their papers by no means always bore his name.

He did have weaknesses...

Of course, Frumkin had weaknesses, too. A principle one was that he did not try to expand his basic knowledge of physics and chemistry, which had remained essentially at a 1920-1930 level. He occasionally did lip service to Eyring's theories (he has known Eyring when spending a year at Madison in the United States in 1932). but he regarded quantum mechanics and statistical mechanics as hot-house toys without much meaning for 'real science'.

Frumkin possessed a psychological weakness, in which all his colleagues the world over were goodies or baddies. Moreover, your goodie-ness or baddie-ness depended largely on how you treated him and his works in your publications. Another Frumkin weakness was his limited ability to discuss a topic calmly, if he had a position in it, but always wanted to make a fight about it, to take sides. 'A tiger of the literature' describes this aspect of Frumkin's approach. Within the Soviet bloc, disagreement with Frumkin could have deleterious results for ones career.

My Reminiscence

My picture is one-might-say a restricted and affectionate reminiscence, containing the warts as well . But let it be said, and the majority of the world's electrochemists would agree, that we have had in Academician Frumkin a great man, who shone like a sun in Moscow for fifty years, and who lead and organized a substantial fraction of the research which illuminated the morning of a field of science, whose noon is yet to come.

**Reprinted in part from an article by Professor Bockris in the Australian Chemical Society Proceedings, Jan. 1977.*

INDEX

Absorbers: 100

Asymmetric Charging, 104

Biographical Names:
- Andre, 68, 98, 99
- Drumm, 161
- Dun, 182
- Edison, 182
- Faraday,6
- Frumkin, 5, 10, 57, 253, 255
- Haschlacher, 182
- Jungner, 161,182
- Ostwald, 10
- Volta, 5

Bipolar Cells, 139

Cellophane, 68, 82, 83, 85

Chemical Equilibrium Equations:
- Debye-Huckel, 38
- Gibbs, 37, 38
- Gibbs-Helmholtz, 7
- Langmuir, 7
- Nernst, 7
- Van't Hoff, 7

Contact Potential, 5

Corrosion, 10, 50, 153, 183
- Background names; 10
  - Davy, de la Rive, Hammett, Lorch, Wagner, Traub
- Pourbaix Diagram, 42, 46, 47
- Simulation, 36
- Tafel Plot, 153
- Test Cell, 53

Current Collectors, 50
- Alloy, 51
- Graphite, 50

Diffusion Effects, 160, 173,175

Electrocapillarity, 7

Electrochemical Kinetics, 8
    Catalysis, 135
    Erdey-Gruz, 8
    Hydrogen evolution, 10
    Levich Plot, 162
    Tafel Equation, 8, 153
    Transfer coefficient, 10
    Volmer, 8

Electrochemical Double Layer, 5
    Gouy-Chapman-Stern, 9
    Frumkin Correction, 9

Electrolytes,
    Additives, 14, 63, 101, 135, 136, 220
    Chemical species in Water-KOH-Zinc solution, 40
    Conductivity, 224
    Modified, 14
    Potassium Hydroxide, 12
    Potassium Fluoride, 135
    Russian, 63
    Solid Polymer, 182
    Thermodynamics, 40

Fuel Cells, 10

Impedance [Impedance Spectroscopy], 14, 195, 204

Manganese Dioxide-Zinc Batteries, 3, 143
    Construction, 116
    Performance, 119
    RAM Cells, 109, 121

Nickel-Electrodes; 172, 210, 241
    charge acceptance, 172
    preparation, 230, 243

Nickel-Zinc Batteries, 3, 160, 182, 195, 210, 220, 229
    Charge acceptance, 172
    Charge retention, 222
    Drumm, 161
    Eectrolyte additives, 222
    Gas recombination, 229
    Polymer electrolyte, 182
    Specific energy, 170

Oxidation of Organics, 10

Oxygen Electrodes, 135, 143
    Catalysts, 135
    Cyclic voltammagram, 154

Porous Electrodes, 5, 11, 124, 136

Rotational Effects on Zinc, 175

Separators,
- Cellulose chemistry, 68
- Cellophane, 63, 68, 99, 106,
  - properties, 71, 82
  - structure, 82
- Electrical resistance, 106
- Electro Permeable Membranes [EPM] 86
- Grafted polyethylene, 9,100
- Microporous polypropylene [Celgard], 92
- Modified cellolosic, 69, 85

Silver electrodes, 101
- Additives, 58
- Sintering, 101

Silver-Zinc Batteries, 3, 57, 68, 84
- Cycling and cycle life, 60, 63, 84, 96
- Design, 84, 98
- Electrolyte, 101, 228
- Energy density, 84, 108
- Power density, 84
- Russian, 61, 103
- Separators, 87, 99
- Utilization, 103,108
- Voltage spikes, 87

Simulation and Modeling, 36

Trade Names:
- AER, 148
- Alkaline Batteries Ltd.,161
- BTI [Battery Technology, Inc], 121
- Celgard, 92,96
- ERC [Energy Research Corporation], 169
- Noranda, 22
- Pure Energy, 109
- Quant Enterprise [Russia], 57
- RAM, 109
- Rayovac, 121
- Renewal, 109
- Sorapec, 135
- Vibrocel, 163,178
- Yardney, 84

Zinc-Air Batteries, 135, 147
    Bipolar, 139
    Dual AIr electrode, 148

Zinc and Zinc Alloys,
    Battery Systems, 3
    Chemical properties, 127
    Chemical Species in Alkaline Electrolyte, 40
    Solubility, 39
    Zincate complexes, 33

Zinc deposition, 15
    Mossy, 14, 16

Zinc Electrodes
    Additives, 13, 22, 87, 183, 232
    Acetylene black effect, 162
    Advantages, 2
    Binders, 125
    Bonded, 195
    Diffusion effects, 103
    Discharge curve, 30
    Equivalent circuit, 197
    Fillers, 124
    Foam, 140
    Hydrogen Evolution, 54
    Modelling, 204
    Passivation, 12, 210, 218
    Potential, 41
    Preparation, 86, 102, 137
        electroformed, 86, 102
        gelled, 122
        pasting, 14, 102, 138
        plastic bonded, 132
        pressed powder, 102
    Production levels, 1
    Reactions, 15, 161
    Solubility, 12
    Shape change, 12, 61, 82
    Utilization, 19
    Vibration effects, 161, 163